Eureka!

Table of Contents

Reader's Guide vii

Inventions and Discoveries by Subject ix

Picture Credits xix

Volume 1: A-B 1

Index 195

Volume 2: C-E 225

Index 413

Volume 3: F-J 443

Index 603

Volume 4: K-O 633

Index 775

Volume 5: P-So 805

Index 951

Volume 6: Sp-Z 981

Index 1179

Reader's Guide

Eureka! Scientific Discoveries and Inventions That Shaped the World features 600 entries on scientific inventions and discoveries that have made a great impact on the world—from the principle of buoyancy to the atomic bomb, from blood transfusion to microcomputers—and the people responsible for them. Written in nontechnical language, *Eureka!* explores such important inventions as the ancient craft of brick-making, but focuses primarily on significant breakthroughs from the Industrial Revolution to the present day, including the invention of the steam engine and the discoveries made possible by the Hubble Space Telescope.

Each *Eureka!* entry, whether on a well-known discovery or a lesser-known invention, identifies the person behind the breakthrough, the knowledge and technology that led to it, and how these advances changed the world in which we live.

Scope and Format

Eureka!'s 600 entries are arranged alphabetically over six volumes. Entries range from one-quarter to eight pages and often include sidebar boxes discussing important breakthroughs, such as firsts in space flight, and lesser-known facts, such as how a frog helped invent the electric battery. Boldfaced terms in the text direct the reader to related entries in the set, while cross-references at the ends of entries alert the reader to related entries not specifically mentioned in that entry. More than 430 photographs and original illustrations enliven and help explain the text.

Each *Eureka!* volume begins with a listing of the featured discoveries and inventions arranged by 37 scientific categories. This handy category listing lets users quickly identify and locate related discoveries and inventions. The comprehensive general index found at the end of each volume provides easy access to the people, theories, and discoveries and inventions mentioned throughout *Eureka!*

Special Thanks and Dedication

The editors dedicate this work to their husbands and to their daughters, Margie and Sara, along with all the children of St. Agatha School in Redford, Michigan, and Bingham Farms Elementary School in Bingham Farms, Michigan. They would also like to express their sincere appreciation to teacher Marlene Heitmanis and tutor and counselor Theresa McCall for their guidance on school curricula.

Comments and Suggestions

We welcome your comments on this work as well as your suggestions for topics to be featured in future editions of *Eureka! Scientific Discoveries and Inventions That Shaped the World.* Please write: Editors, *Eureka!* U·X·L, 835 Penobscot Bldg., Detroit, Michigan 48226-4094; call toll-free: 1-800-877-4253; or fax 1-313-877-6348.

Inventions and Discoveries by Subject

Bold numerals indicate volume numbers.

Agriculture

Animal breeding **1:** 64
Barbed wire **1:** 133
Canal and canal lock **2:** 231
Fertilizer, synthetic **3:** 453
Hydroponics **3:** 569

Astronomy

Big bang theory **1:** 147
Binary star **1:** 154
Earth's magnetic field **2:** 370
Einstein, Albert **2:** 374
Electromagnetic wave **2:** 385
Galileo Galilei **3:** 482
Gamma ray astronomy **3:** 487
Hawking, Stephen William **3:** 526
Jupiter **3:** 597
Mars **4:** 676
Mercury (planet) **4:** 686
Meteor and meteorite **4:** 689
Moon **4:** 709
Neptune **4:** 728
Newton, Isaac **4:** 737
Nova and supernova **4:** 745
Planetary motion **5:** 838
Plasma **5:** 841
Pluto **5:** 845
Pulsar **5:** 863
Quasar **5:** 868
Radio astronomy **5:** 879
Red-shift **5:** 886
Sagan, Carl **5:** 910
Saturn **5:** 913
Solar system **5:** 943
Solar wind **5:** 947
Star cluster **6:** 1010
Steady-state theory **6:** 1012
Sun **6:** 1031
Ultraviolet astronomy **6:** 1111
Uranus **6:** 1118
Van Allen belts **6:** 1125
Venus **6:** 1127
X-ray astronomy **6:** 1162

Automotive engineering

Airbag, automobile **1:** 26
Automobile, electric **1:** 118
Automobile, gasoline **1:** 121
Cruise control, automobile **2:** 329
Internal combustion engine **3:** 584
Windshield and windshield wiper **6:** 1152

Aviation and aerospace

Aircraft **1:** 29
Altimeter **1:** 46
Automatic pilot **1:** 117
Communications satellite **2:** 286
Earth survey satellite **2:** 372
Explorer 1 **2:** 406
Helicopter **3:** 535
Jet engine **3:** 595
Kite **4:** 635

Navigational satellite **4:** 725
Rocket and missile **5:** 900
Solar sail **5:** 941
Spacecraft, manned **6:** 981
Space equipment **6:** 988
Space probe **6:** 994
Space shuttle **6:** 997
Space station **6:** 1000
Stealth aircraft **6:** 1013
Wind tunnel **6:** 1155

Biochemistry

Acetylcholine **1:** 4
ACTH (adrenocorticotropic hormone) **1:** 14
Amino acid **1:** 52
Cholesterol **2:** 269
Cortisone **2:** 322
DNA (deoxyribonucleic acid) **2:** 352
Endorphin and enkephalin **2:** 393
Enzyme **2:** 398
Epinephrine (adrenaline) **2:** 399
Fermentation **3:** 450
Glycerol **3:** 509
Hormone **3:** 552
Human growth hormone (somatotropin) **3:** 560
Lipid **4:** 656
Nerve growth factor **4:** 730
Niacin **4:** 739
Phagocyte **5:** 824
Restriction enzyme **5:** 890
Secretin **5:** 914
TRH (thyrotropin-releasing hormone) **6:** 1094
Urea **6:** 1120
Vitamin **6:** 1136

Biology

AIDS (Acquired Immune Deficiency Syndrome) **1:** 20
Antibody and antigen **1:** 71
Bacteria **1:** 128
Blood **1:** 165
Cell **2:** 258
Chromosome **2:** 269
Cloning **2:** 275
Digestion **2:** 345
DNA (deoxyribonucleic acid) **2:** 352
Evolutionary theory **2:** 401
Fertilization **3:** 453
Gene **3:** 493
Genetic code **3:** 498
Germ theory **3:** 504
Heredity **3:** 543
Human evolution **3:** 554
Inoculation **3:** 575
Lymphatic system **4:** 662
Metabolism **4:** 687
Mutation **4:** 720
Nervous system **4:** 731
Neuron theory **4:** 733
Nitrogen **4:** 741
Pavlov, Ivan Petrovich **5:** 812
Photosynthesis **5:** 835
Population genetics **5:** 855
Rh factor **5:** 891
RNA (ribonucleic acid) **5:** 893
Sex chromosome **5:** 919
Split-brain functioning **6:** 1008
Synapse **6:** 1038
Tissue **6:** 1069
Virus **6:** 1133
Yeast **6:** 1169

Chemistry

Acetic acid **1:** 3
Acid and base **1:** 6
Alcohol, distilling of **1:** 37
Allotropy **1:** 40
Artificial sweetener **1:** 93
Atomic and molecular weights **1:** 105
Bleach **1:** 161
Boyle's law **1:** 180
Bromine **1:** 189
Bunsen burner **1:** 190
Carbon **2:** 238
Carbon dioxide **2:** 241
Carbon monoxide **2:** 243
Combustion **2:** 282
Cryogenics **2:** 330
Curium **2:** 331
DDT (dichloro-diphenyl-trichloroethane) **2:** 336

Desalination techniques **2:** 342
Dry cleaning **2:** 357
Dynamite **2:** 359
Einsteinium **2:** 377
Electrolysis **2:** 384
Elements 104-109 **2:** 390
Ethylene **2:** 400
Fats and oils **3:** 447
Fluorine **3:** 470
Folic acid **3:** 471
Gasoline **3:** 491
Gold **3:** 509
Helium **3:** 537
Hydrocarbon **3:** 561
Hydrochloric acid **3:** 562
Hydrogen **3:** 564
Isomer **3:** 593
Isotope **3:** 594
Krypton **4:** 637
Lead **4:** 645
Lipid **4:** 656
Magnesium **4:** 667
Manganese **4:** 674
Mercury (element) **4:** 684
Methane **4:** 694
Molecular structure **4:** 707
Natural gas **4:** 723
Neon **4:** 726
Neptunium **4:** 729
Nitric acid **4:** 740
Nitrogen **4:** 741
Nylon **4:** 755
Oil refining **4:** 759
Osmium **4:** 765
Osmosis **4:** 766
Oxygen **4:** 767
Ozone **4:** 770
Pauling, Linus **5:** 810
Periodic law **5:** 818
Petroleum **5:** 821
Phosphorus **5:** 825
Photochemistry **5:** 826
Plastic **5:** 842
Platinum **5:** 844
Polychlorinated biphenyls (PCBs) **5:** 850
Polyester **5:** 851
Polyethylene **5:** 852
Polymer and polymerization **5:** 853
Polypropylene **5:** 854
Polystyrene **5:** 854
Polyurethane **5:** 855
Potassium **5:** 856
Protactinium **5:** 860
Radioactive tracer **5:** 876
Radioactivity **5:** 877
Radium **5:** 881
Radon **5:** 882
Samarium **5:** 912
Selenium **5:** 918
Silicon **5:** 923
Silver **5:** 928
Sodium **5:** 939
Technetium **6:** 1048
Thallium **6:** 1064
Tin **6:** 1068
Titanium **6:** 1070
Transmutation of elements **6:** 1088
Tungsten **6:** 1095
Uranium **6:** 1117
Xenon **6:** 1159
Valence **6:** 1124

Civil engineering and construction

Arch **1:** 74
Brick **1:** 184
Bridge **1:** 186
Canal and canal lock **2:** 231
Earthquake-proofing techniques **2:** 366
Elevator **2:** 391
Geodesic dome **3:** 502
Iron **3:** 591
Road building **5:** 893
Skyscraper **5:** 931
Tunnel **6:** 1095

Clothing and textiles and their manufacture

Blue jeans **1:** 178
Buttons and other fasteners **1:** 193
Carpet **2:** 250
Fiber, synthetic **3:** 459
Nylon **4:** 755
Polyester **5:** 851

Velcro **6:** 1126
Zipper **6:** 1176

Communications/graphic arts

Alphabet **1:** 41
Audiocassette **1:** 114
Blind, communication systems for **1:** 163
Cable television **2:** 225
Cathode-ray tube **2:** 256
Color photography **2:** 278
Communications satellite **2:** 286
Compact disc player **2:** 289
Fax machine **3:** 449
Fiber optics **3:** 456
High-speed flash photography **3:** 548
Instant camera (Polaroid Land camera) **3:** 579
Language, universal **4:** 639
Magnetic recording **4:** 668
Microphone **4:** 697
Microwave transmission **4:** 702
Motion picture **4:** 713
Movie camera **4:** 715
Noise reduction system **4:** 744
Pen **5:** 814
Photocopying **5:** 827
Photography **5:** 830
Phototypesetting **5:** 836
Post-it note **5:** 856
Radar **5:** 869
Radio **5:** 874
Shortwave radio **5:** 920
Sign language **5:** 921
Silk-screen printing **5:** 926
Skywriting **5:** 932
Stereo **6:** 1019
Telephone answering device **6:** 1048
Telephone cable, transatlantic **6:** 1049
Teleprinter and teletype **6:** 1051
Television **6:** 1056
3-D motion picture **6:** 1065
Typewriter **6:** 1101
Underwater photography **6:** 1115
Voice synthesizer **6:** 1138
Walkman **6:** 1141

Computer science and mathematical devices

Artificial intelligence (AI) **1:** 86
Calculator, pocket **2:** 228
Cash register **2:** 253
Computer, analog **2:** 293
Computer application **2:** 294
Computer, digital **2:** 298
Computer disk and tape **2:** 303
Computer, industrial uses of **2:** 305
Computer input and output device **2:** 308
Computer network **2:** 310
Computer operating system **2:** 311
Computer simulation **2:** 313
Computer speech recognition **2:** 315
Computer vision **2:** 316
Cybernetics **2:** 332
Digitizer **2:** 348
Expert system **2:** 405
Machine language **4:** 665
Mainframe computer **4:** 671
Microcomputer **4:** 695
Microprocessor **4:** 699
Minicomputer **4:** 705
Optical disk **4:** 764
Robotics **5:** 897
Supercomputer **6:** 1034
Teleprinter and teletype **6:** 1051
Video game **6:** 1129
Voice synthesizer **6:** 1138

Electrical engineering/electricity

Alternating current **1:** 44
Battery, electric **1:** 143
Cathode-ray tube **2:** 256
Neon light **4:** 727
Polychlorinated biphenyls (PCBs) **5:** 850
Telephone answering device **6:** 1048
Telephone cable, transatlantic **6:** 1049
Teleprinter and teletype **6:** 1051
Television **6:** 1056

Electronics

Amplifier **1:** 54
Audiocassette **1:** 114
Calculator, pocket **2:** 228
Fax machine **3:** 449
LCD (liquid crystal display) **4:** 644
LED (light-emitting diode) **4:** 645
Magnetic recording **4:** 668
Microphone **4:** 697
Noise reduction system **4:** 744
Oscillator **4:** 765
Radio **5:** 874
Video recording **6:** 1130
Walkman **6:** 1141

Environmental sciences/ecology

Acid rain **1:** 9
Food chain **3:** 471
Gaia hypothesis **3:** 481
Greenhouse effect **3:** 513
Methane **4:** 694
Natural gas **4:** 723
Ozone **4:** 770
Radon **5:** 882
Recycling **5:** 884
Red tide **5:** 887

Everyday items

Aerosol spray **1:** 19
Baby bottle **1:** 127
Baby carrier/pouch **1:** 127
Bar code **1:** 135
Bath and shower **1:** 140
Calculator, pocket **2:** 228
Can opener **2:** 237
Car wash, automatic **2:** 252
Cash register **2:** 253
Chewing gum **2:** 264
Diaper, disposable **2:** 344
Dog biscuit **2:** 356
Electric blanket **2:** 381
Eraser **2:** 400
Fan **3:** 445
Fire extinguisher **3:** 462
Firefighting equipment **3:** 462
Flashlight **3:** 467
Instant coffee **3:** 580
Laundromat **4:** 642
Pen **5:** 814
Polystyrene **5:** 854
Post-it note **5:** 856
Stapler **6:** 1009
Swiss army knife **6:** 1038
Teaching aid **6:** 1045
Telephone answering device **6:** 1048
Toilet **6:** 1072
Toothbrush and toothpaste **6:** 1074
Traffic signal **6:** 1079
Tupperware **6:** 1100
Typing correction fluid **6:** 1104
Umbrella **6:** 1114
Vacuum bottle **6:** 1123
Waterbed **6:** 1143

Food/food science

Artificial sweetener **1:** 93
Baby food, commercial **1:** 128
Bread and crackers **1:** 180
Breakfast cereal **1:** 182
Can and canned food **2:** 234
Can opener **2:** 237
Chewing gum **2:** 264
Chocolate **2:** 266
Concentrated fruit juice **2:** 317
Doughnut **2:** 356
Fat substitute **3:** 448
Food preservation **3:** 472
Soda pop **5:** 938
Yogurt **6:** 1173

Geology

Continental drift **2:** 319
Cretaceous catastrophe **2:** 326
Earthquake **2:** 363
Earthquake measurement scale **2:** 365
Earth's core **2:** 368
Earth's mantle **2:** 371
Earth survey satellite **2:** 372
Mohorovicic discontinuity **4:** 705
Plate tectonics **5:** 843
Seismology **5:** 916
Uniformitarianism **6:** 1116

Household appliances

Clothes dryer **2:** 277
Dishwasher **2:** 349
Lawn mower **4:** 643
Microwave oven **4:** 700
Toaster **6:** 1071
Washing machine **6:** 1141

Lighting and illumination

Halogen lamp **3:** 523
Neon light **4:** 727

Materials

Abrasive **1:** 1
Acrylic plastic **1:** 13
Adhesives and adhesive tape **1:** 17
Biodegradable plastic **1:** 155
Dacron **2:** 335
Fiberglass **3:** 456
Kevlar **4:** 633
Recycling **5:** 884
Rubber, vulcanized **5:** 905
Saran **5:** 913
Silicone **5:** 924
Vinyl **6:** 1132
Waterproof material **6:** 1143

Mathematics

Algorithm **1:** 38
Binary arithmetic **1:** 151
Binomial theorem **1:** 154
Calculable function **2:** 227
Decimal system **2:** 338
Game theory **3:** 485
Irrational number **3:** 593
Logarithm **4:** 660
Newton, Isaac **4:** 737
Real number **5:** 883
Roman numerals **5:** 904
Topology **6:** 1075
Zero **6:** 1175

Mechanical engineering

Air conditioning **1:** 27
Cybernetics **2:** 332
Engine **2:** 394
Engine oil **2:** 396
Heating **3:** 532
Ice-resurfacing machine **3:** 572
Internal combustion engine **3:** 584
Mass production **4:** 678
Photocopying **5:** 827
Robotics **5:** 897
Steam engine **6:** 1016

Medicine/dentistry/health sciences

Acoustics, physiological **1:** 12
Acupuncture **1:** 15
Allergy **1:** 39
Ambulance **1:** 50
Amniocentesis **1:** 53
Amputation **1:** 56
Anemia **1:** 57
Anesthesia **1:** 59
Angioplasty, balloon **1:** 61
Arteriosclerosis **1:** 82
Artificial heart **1:** 84
Artificial ligament **1:** 89
Artificial limb and joint **1:** 90
Artificial skin **1:** 91
Audiometer **1:** 115
Bandage and dressing **1:** 132
Birth control **1:** 158
Blood **1:** 165
Blood, artificial **1:** 170
Blood clot dissolving agent **1:** 171
Blood pressure measuring device **1:** 171
Blood transfusion **1:** 173
Blood vessels, artificial **1:** 177
Carcinogen **2:** 245
Cataract surgery **2:** 254
Catheter, cardiac **2:** 256
Cholera **2:** 268
Cholesterol **2:** 269
Chromosome **2:** 269
Cocaine **2:** 277
Contact lens **2:** 318
Cortisone **2:** 322
Cystic fibrosis **2:** 333
Dental drill **2:** 340

Dental filling, crown, and bridge **2:** 341
Dialysis machine **2:** 343
Digestion **2:** 345
Electrocardiograph (ECG) **2:** 382
Electroencephalogram (EEG) **2:** 383
Endorphin and enkephalin **2:** 393
Endoscope **2:** 393
Eye disorders **2:** 407
Eyeglasses **2:** 410
False teeth **3:** 443
Fluoride treatment, dental **3:** 468
Folic acid **3:** 471
Fractures, treatments and devices for treating **3:** 478
Gene therapy **3:** 495
Genetically engineered blood-clotting factor **3:** 497
Genetic engineering **3:** 499
Genetic fingerprinting **3:** 501
Germ theory **3:** 504
Glaucoma **3:** 508
Gonorrhea **3:** 510
Hearing aids and implants **3:** 528
Heart-lung machine **3:** 529
Hemophilia **3:** 539
Hepatitis **3:** 541
Hormone **3:** 552
Human growth hormone (somatotropin) **3:** 560
Hypertension **3:** 570
Ibuprofen **3:** 571
Immune system **3:** 573
Inoculation **3:** 575
Insulin **3:** 581
Interferon **3:** 583
In vitro fertilization **3:** 587
Lyme disease **4:** 661
Lymphatic system **4:** 662
Magnetic resonance imaging (MRI) **4:** 670
Mammography **4:** 673
Meningitis **4:** 682
Morphine **4:** 713
Mumps **4:** 717
Muscular dystrophy **4:** 717
Nerve growth factor **4:** 730
Nervous system **4:** 731
Neuron theory **4:** 733
Novocain **4:** 747
Nuclear magnetic resonance (NMR) **4:** 751
Pacemaker **5:** 805
Pap test **5:** 806
Pauling, Linus **5:** 810
PKU (phenylketonuria) **5:** 837
Pneumonia **5:** 847
Polio **5:** 849
Prenatal surgery **5:** 859
Radial keratotomy **5:** 872
Radioactive tracer **5:** 876
Radiotherapy **5:** 881
RU 486 **5:** 907
Sleeping sickness **5:** 933
Syphilis **6:** 1041
Teratogen **6:** 1060
Tetanus **6:** 1063
Tissue **6:** 1069
Toothbrush and toothpaste **6:** 1074
Transplant, surgical **6:** 1090
Ultrasound device **6:** 1108
X-ray machine **6:** 1165
Yellow fever **6:** 1171

Metallurgy

Alloy **1:** 40
Aluminum production **1:** 48

Meteorology

Atmospheric circulation **1:** 98
Atmospheric composition and structure **1:** 101
Atmospheric pressure **1:** 104
Ionosphere **3:** 588
Meteor and meteorite **4:** 689
Meteorology **4:** 692
Storm **6:** 1020
Weather forecasting model **6:** 1145
Weather satellite **6:** 1149

Musical instruments

Musical instrument, electric **4:** 718
Synthesizer, music **6:** 1040

Navigation

Astrolabe **1:** 96
Buoy **1:** 190
Compass **2:** 290
Navigational satellite **4:** 725
Sonar **5:** 949

Oceanography

Decompression sickness **2:** 339
Diving apparatus **2:** 350
Mid-ocean ridges and rifts **4:** 704
Oceanography **4:** 757

Optics

Contact lenses **2:** 318
Endoscope **2:** 393
Eyeglasses **2:** 410
Hologram **3:** 549
Kaleidoscope **4:** 633
Laser **4:** 640
Lens **4:** 647
Solar telescope **5:** 946
Space telescope **6:** 1003
Telescope **6:** 1052

Personal care items

Cosmetics **2:** 322
Hair care **3:** 519
Toothbrush and toothpaste **6:** 1074

Pharmacology

AIDS therapies and vaccines **1:** 24
Antibiotic **1:** 66
Barbiturate **1:** 135
Chemotherapy **2:** 263
Cocaine **2:** 277
Hallucinogen **3:** 521
Heroin **3:** 547
Ibuprofen **3:** 571
Interferon **3:** 583
Morphine **4:** 713
Novocain **4:** 747
Penicillin **5:** 816
Sulfonamide drug **6:** 1030
Teratogen **6:** 1060
Tranquilizer (antipsychotic type) **6:** 1086

Physics

Absolute zero **1:** 2
Antiparticle **1:** 72
Atomic theory **1:** 112
Buoyancy, principle of **1:** 191
Cathode-ray tube **2:** 256
Color spectrum **2:** 280
Cryogenics **2:** 330
Einstein, Albert **2:** 374
Elasticity **2:** 378
Electric arc **2:** 380
Electric charge **2:** 381
Electromagnetic wave **2:** 385
Electromagnetism **2:** 386
Electron **2:** 388
Fiber optics **3:** 456
Fluorescence and phosphorescence **3:** 467
Galileo Galilei **3:** 482
Gamma ray **3:** 486
Gravity **3:** 511
Hawking, Stephen William **3:** 526
Heat and thermodynamics **3:** 530
High-pressure physics **3:** 547
Kinetic theory of gases **4:** 634
Light, diffraction of **4:** 650
Light, polarization of **4:** 650
Light, reflection and refraction of **4:** 651
Light, theories of **4:** 653
Magnetic field **4:** 668
Mass spectrograph **4:** 681
Metal fatigue **4:** 688
Microwave **4:** 700
Neutron **4:** 734
Neutron bomb **4:** 736
Newton, Isaac **4:** 737
Nuclear fission **4:** 748
Nuclear fusion **4:** 749
Nuclear magnetic resonance (NMR) **4:** 751
Nuclear reactor **4:** 753
Particle accelerator **5:** 807
Particle spin **5:** 810
Photoelectric effect **5:** 828

Piezoelectric effect **5:** 836
Plasma **5:** 841
Proton **5:** 861
Quantum mechanics **5:** 865
Quantum theory **5:** 867
Radiation detector **5:** 872
Radio interferometer **5:** 880
Red-shift **5:** 886
Relativity **5:** 888
Spectroscopy **6:** 1006
Subatomic particle **6:** 1022
Superconductivity **6:** 1035
Tunneling **6:** 1099
Ultrasonic wave **6:** 1107
Ultraviolet radiation **6:** 1112
Wave motion, law of **6:** 1144
X-ray **6:** 1160
X-ray machine **6:** 1165

Security systems and related items

Fingerprinting **3:** 460
Lie detector **4:** 649
Lock and key **4:** 657
Retinography **5:** 891
Safe **5:** 909

Sports, games, toys, and fads

Baseball **1:** 137
Basketball **1:** 138
Comic strip and comic book **2:** 284
Ferris wheel **3:** 451
Fireworks **3:** 465
Football **3:** 476
Frisbee **3:** 479
Hula hoop **3:** 554
Rubik's cube **5:** 907
Silly Putty **5:** 927
Skateboard **5:** 929
Skis and ski bindings **5:** 930
Slinky **5:** 934
Slot machine and vending machine **5:** 935
Surfboard **6:** 1036
Trampoline **6:** 1085
Video game **6:** 1129
Windsurfer **6:** 1154

Timepieces, measuring devices, and related items

Astrolabe **1:** 96
Atomic clock **1:** 111
Clock and watch **2:** 271
Time zone **6:** 1066

Transportation

Ambulance **1:** 50
Cargo ship **2:** 248
Elevator **2:** 391
Helicopter **3:** 535
Hydrofoil **3:** 563
Snowmobile **5:** 937
Submarine **6:** 1024
Subway **6:** 1028
Train and railroad **6:** 1080
Wheelchair **6:** 1151

Weapons and related items

Armor **1:** 77
Armored vehicle **1:** 80
Artillery **1:** 94
Atomic bomb **1:** 107
Bayonet **1:** 145
Bazooka **1:** 146
Biological warfare **1:** 156
Chemical warfare **2:** 260
Flamethrower **3:** 466
Gas mask **3:** 488
Grenade **3:** 516
Gun silencer **3:** 517
Harpoon **3:** 524
Hydrogen bomb **3:** 567
Neutron bomb **4:** 736
Rocket and missile **5:** 900
Torpedo **6:** 1077

Picture Credits

The photographs appearing in *Eureka! Scientific Discoveries and Inventions That Shaped the World* were received from the following sources:

©Bernard Uhlig/Phototake NYC: p. 9; **©Yoav Levy/Phototake NYC:** pp. 16, 392, 549, 806, 819, 931; **©Tony Freeman/Photo Edit:** pp. 18, 325, 380, 462, 543, 562, 638; **AP/Wide World Photos:** pp. 21, 30, 35, 67, 68, 78, 124, 139, 141, 151, 164, 257, 261, 266, 285, 294, 298, 314, 332, 365, 373, 374, 465, 495, 516, 523, 578, 580, 672, 676, 679, 695, 696, 714, 731, 747, 748, 749, 823, 830, 833, 850, 852, 854, 865, 866, 888, 902, 910, 917, 930, 938, 945, 984, 994, 1002, 1008, 1020, 1034, 1044, 1047, 1057, 1059, 1061, 1079, 1090, 1093, 1153, 1172; **UPI/Bettmann:** pp. 26, 32, 51, 79, 90, 137, 148, 159, 235, 272, 287, 289, 364, 450, 458, 477, 490, 504, 546, 558, 559, 599, 634, 681, 689, 699, 710, 727, 807, 811, 849, 871, 901, 905, 911, 934, 1027, 1029, 1046, 1077, 1150, 1176; **Reuters/Bettmann:** pp. 37, 81, 601, 878, 916; **Bettmann Archive:** pp. 42, 43, 121, 122, 145, 157, 174, 180, 185, 228, 244, 253, 401, 411, 445, 484, 489, 533, 566, 582, 677, 690, 716, 733, 813, 816, 831, 832, 839, 862, 867, 875, 876, 882, 913, 919, 939, 948, 983, 989, 1016, 1018, 1038, 1042, 1053, 1063, 1102, 1136, 1167; **©Michael Newman/Photo Edit:** pp. 49, 448, 853, 1096; **©The Telegraph Colour Library/FPG International:** pp. 59, 100, 226, 320, 598; **©Anna E. Zuckerman/Photo Edit:** p. 64; **©Jon Gordon/Phototake NYC:** p. 65; **©Kent Knudson/FPG International:** p. 76; **©Martin Roiker/Phototake NYC:** p. 82; **©Account Phototake/Phototake NYC:** pp. 85, 108, 501, 742; **©Howard Sochurek 1986/The Stock Market:** p. 87; **Dar al-Athar al-Islamiyyah, Ministry of Information, Kuwait:** p. 97; **Courtesy of the Department of Energy:** p. 119; **©1992 Howard Sochurek/The Stock Market:** p. 129; **©Dr. Den-**

nis Kunkel/Phototake NYC: pp. 130, 260, 1039; ©David Young-Wolff/Photo Edit: pp. 136, 378, 449; The Science Museum/Science & Society Picture Library: pp. 143, 274; ©1976 Isaiah Karlinsky/FPG International: p. 163; ©Goivaux Communication/Phototake NYC: p. 165; ©1988 David Frazier/The Stock Market: p. 172; ©1986 Randy Duchaine/The Stock Market: p. 178; ©Hammond Incorporated, Maplewood, New Jersey, License #12, 231: 187; ©Roy Morsch 1983/The Stock Market: p. 191; ©Felicia Martinez/Photo Edit: p. 193; ©Ron Routar 1991/FPG International: p. 230; Peter L. Gould/FPG International: p. 233; ©Michael Simpson 1991/ FPG International: p. 234; Arizona Historical Society Library: p. 247; ©Deborah Davis/Photo Edit: p. 249; ©Margaret Cubberly/Phototake NYC: p. 255; ©Earl Young/FPG International: p. 262; ©CNRI/Phototake NYC: p. 270; Phototake: pp. 281, 655, 835; ©NASA/SB/FPG International: p. 288; The Granger Collection, New York: p. 292; ©1991 Brownie Harris/The Stock Market: p. 296; UPI/Bettmann Newsphotos: pp. 300, 847; ©1985 Paul Ambrose/FPG International: p. 304; ©Tom Carroll 1988/FPG International: p. 306; ©Spencer Grant/FPG International: pp. 309, 586; FPG International: p. 323; ©Frank Rossotto/The Stock Market: pp. 328, 925; Archive Photos/Orville Logan Snider: p. 337; Magnum Photos: p. 338; ©Dennis Kunkel/CNRI/Phototake NYC: p. 347; ©Art Montes De Oca 1989/ FPG International: p. 348; Mary Evans Picture Library: pp. 351, 1025; ©Dr. Louise Chow/Phototake NYC: p. 353; Cold Spring Harbor Laboratory Archives: p. 354; Reuters/Bettmann Newsphotos: p. 367; ©Dr. David Rosenbaum/Phototake NYC: p. 383; Bill Wisser/FPG International: p. 386; National Portrait Gallery, London: p. 387; ©Jeff Greenberg/Photo Edit: p. 395; ©Tom Campbell 1989/FPG International: p. 397; Bob Abraham/The Stock Market: pp. 408, 641; ©Spencer Jones 1993/FPG International: p. 463; Michael Price/FPG International: p. 473; ©Gabe Palmer/Kane, Inc. 1982/The Stock Market: p. 475; Anthony Howarth/Science Photo Library/Photo Researchers, Inc.: p. 482; Thomas Lindsay/FPG International: p. 509; Illustration from *Levitating Trains and Kamikaze Genes: Technological Literacy for the 1990's,* by Richard P. Brennan. Copyright © 1990 by John Wiley & Sons, Inc. Reprinted by permission of John Wiley & Sons, Inc.: p. 513; ©Holt Confer/Phototake NYC: p. 514; ©Miriam Berkley: p. 526; Michael Johnson; courtesy of Green-Peace: p. 534; ©1990 Jack Van Antwerp/The Stock Market: p. 536; E. Bernstein and E. Kairinen, Gillette Research Institute: p. 540; ©Carolina Biological Supp/Phototake NYC: p. 545; National Museum of

Medicine: p. 576; **©1987 Randy Duchaine/The Stock Market:** p. 592; **©1993 John Olson/The Stock Market:** p. 597; **©1988 Don Mason/The Stock Market:** p. 642; **©1991 Ken Korsh/FPG International:** p. 643; **©Granada Studios/FPG International:** p. 644; **©Ron Routar 1991/FPG International:** p. 646; **Arthur Gurman Kin/Phototake NYC:** p. 652; **Tom McCarthy Photos/Photo Edit:** p. 659; **©Robert Reiff 1991/FPG International:** p. 667; **©R. Rathe 1993/FPG International:** p. 671; **©J. Barry O'Rourke/The Stock Market:** p. 673; **©Kent Knudson 1991/FPG International:** p. 675; **©1994 Eugene Smith/Black Star:** p. 686; **NASA:** pp. 687, 728, 769, 903, 986, 991, 992, 999; **Dick Kent Photography/FPG International:** p. 691; **©Frank Rossotto 1992/The Stock Market:** p. 693; **©Jack Zehrt 1992/FPG International:** p. 711; **©Leslye Borden/Photo Edit:** p. 719; **©Peter Vadnai/The Stock Market:** p. 736; **©Overseas/Phototake NYC:** p. 750; **©Paul Ambrose 1988/ FPG International:** p. 751; **Robert Visser; courtesy of GreenPeace:** p. 754; **©Steve McCutcheon/Visuals Unlimited:** p. 758; **©Peter Britton/Phototake NYC:** p. 759; **©Zachary Singer/GreenPeace 1989:** p. 761; **NASA/Phototake:** pp. 772, 1004; **©Mauritus GMBH/Phototake NYC:** p. 808; **©Michael Seigel/Phototake NYC:** p. 818; **LeKarer; courtesy of GreenPeace:** p. 822; **USGS:** p. 844; **John Gajda/FPG International:** p. 857; **©1991 Howard Sochurek/The Stock Market:** p. 859; **©1991 Peter Beck/The Stock Market:** p. 884; **©Robert Ginn/Photo Edit:** p. 894; **©Peter Gridley 1987/FPG International:** p. 896; **©John Madere/The Stock Market:** p. 898; **©Steve Kahn 1990/FPG International:** p. 899; **©Paul Ambrose 1986/FPG International:** p. 924; **Tass/Sovfoto:** p. 988; **©Jeff Divine 1991/FPG International:** p. 1037; **©1994 Peggy and Ronald Barnett/The Stock Market:** p. 1050; **©Movie Still Archives 1994/FPG International:** p. 1065; **©Gabe Palmer/The Stock Market:** p. 1068; **©Phototake Kunkel/Phototake NYC:** p. 1069; **Courtesy of I.G.D. Dunlap:** p. 1072; **Indianapolis & Louisville Ry.:** p. 1081; **UP:** p. 1084; **©AAA Photo/Phototake NYC:** p. 1085; **©Ann Chwatsky/Phototake NYC:** p. 1091; **Courtesy of The Mark Twain Papers, Bancroft Library:** p. 1103; **©Richard Nowitz 1990/ FPG International:** p. 1109; **©Wagner Herbert Stock/Phototake NYC:** p. 1133; **©1988 Jim Brown/The Stock Market:** p. 1146; **©Peter A. Simon/Phototake NYC:** pp. 926, 1155; **Royal Institution:** p. 1161.

The original illustrations appearing in *Eureka!* were researched and drawn by Teresa SanClementi.

Eureka!

F

False teeth

In earlier times, upper and lower plates fit poorly and were held together with steel springs. Many a diner was embarrassed when his teeth suddenly sprang out of his mouth.

Partly for beauty's sake and partly to aid in digestion, replacements for decayed or lost teeth have been produced for thousands of years. As early as 700 B.C., the Etruscans in Italy made skillfully designed false teeth out of ivory and bone, held in place by gold bridgework. Unfortunately, this level of sophistication for false teeth was not regained until the 1800s.

During medieval times (400-1450), the practice of dentistry was largely confined to tooth extraction. Gaps between teeth were expected, even among the rich and powerful. Queen Elizabeth I (1533-1603) of England filled the holes in her mouth with cloth to improve her appearance in public.

When false teeth were installed, they were hand-carved and tied in place with silk threads. People who wore full sets of dentures had to remove them when they wanted to eat. Even George Washington (1732-1799) suffered terribly from tooth loss and ill-fitting dentures. The major obstacles to progress were finding suitable materials for false teeth, making accurate measurements of a patient's mouth, and getting the teeth to stay in place. These problems began to be solved during the 1700s.

Better Materials

Since antiquity, the most common material for false teeth was animal bone or ivory, especially from elephants or hippopotamuses. Human teeth were also used, pulled from the dead or sold by poor people from their own mouths. These kinds of false teeth soon rotted, turning brown and rancid (smelly). Rich people preferred teeth of silver, gold, mother of pearl, or agate (a type of stone).

In 1774 hard-baked, rot-proof porcelain dentures came on the French market. Then the single porcelain tooth held in place by an imbedded platinum pin was invented in 1808 in Italy. An improved porcelain tooth appeared in England around 1837.

Porcelain teeth came to the United States in 1817. Commercial manufacture of porcelain teeth in the United States was begun in Philadelphia around 1825 by Samuel Stockton. In 1844 Stockton's nephew founded the S. S. White Company, which greatly improved the design of artificial teeth and marketed them on a large scale.

Better Fit

Fit and comfort, too, gradually improved. The German Philip Pfaff (1715-1767) introduced plaster of paris impressions of the patient's mouth in 1756. Daniel Evans of Philadelphia also devised a method of accurate mouth measurement in 1836.

The real breakthrough came with Charles Goodyear's discovery of vulcanized rubber in 1839. This cheap, easy-to-work material could be

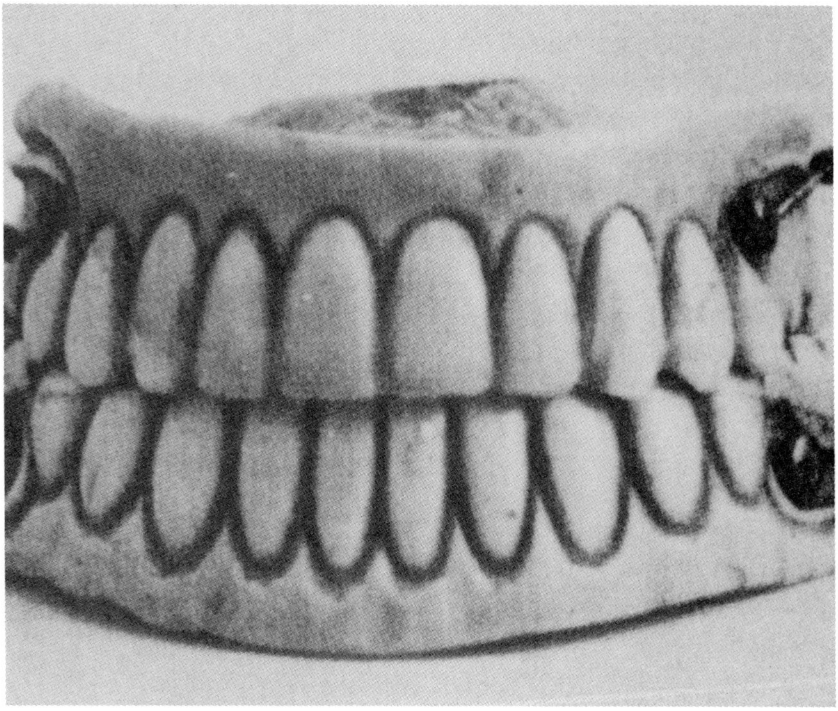

Ivory teeth carved during the early nineteenth century. The Etruscans in Italy made skillfully designed false teeth out of ivory and bone as early as 700 B.C.

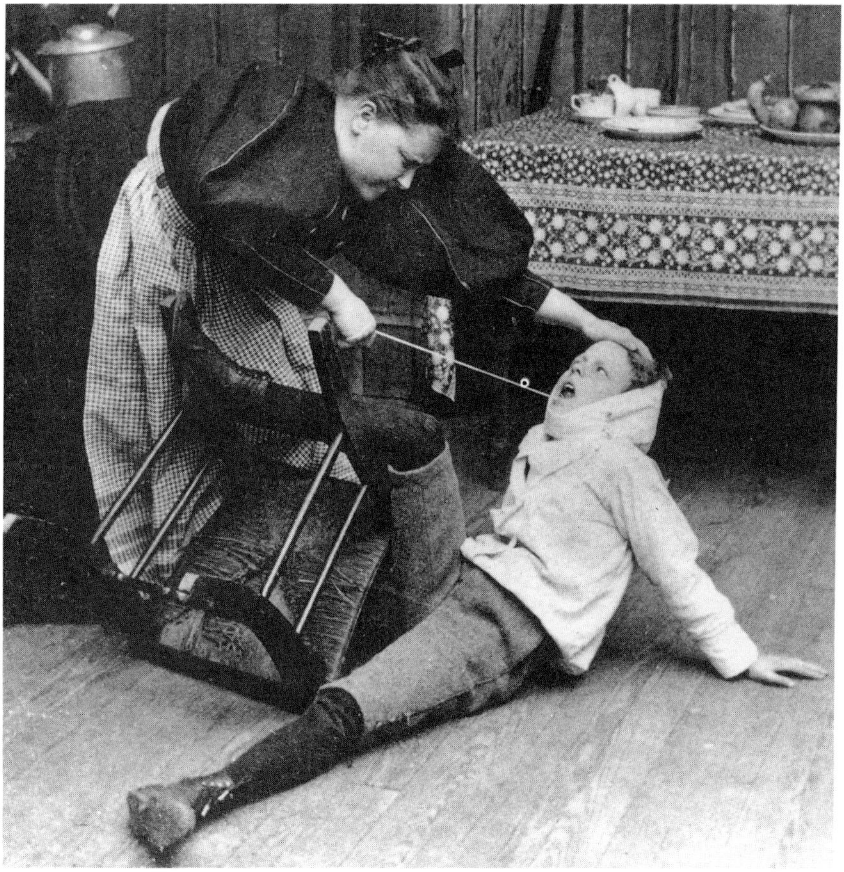

The number of people having teeth removed skyrocketed after tooth extraction was made painless with nitrous oxide. A great demand for good, affordable dentures followed.

molded to fit the mouth and made a good base to hold false teeth. Well-mounted dentures could now be made cheaply.

The timing could not have been better. Horace Wells (1815-1848) had just introduced painless tooth extraction (pulling) using nitrous oxide (laughing gas). The number of people having teeth removed skyrocketed, creating a great demand for good, affordable dentures, which Goodyear's invention made possible. Today dentures are either **plastic** or ceramic.

Fan

Fans have been used for thousands of years as tools for keeping cool and as fashion statements. Egyptian pharaohs were fanned by their slaves with huge lotus leaves. Ancient Greeks and Romans often trimmed their fans with peacock feathers. The fan served both practical and ceremonial functions in China and Japan. In China it was especially popular during the Ming Dynasty (1368-1644). Fans were carried by men and women from members of many different social classes during tea ceremonies and on stage. Some of the most talented Asian painters applied their skill to the exquisite decoration of fans, an art that was not developed in Europe until the nineteenth century.

The folding or pleated fan is thought to have been invented by the Japanese in about 700 A.D. and may have been modeled after the way a bat folds its wings.

Fans in Fashion

During the Middle Ages (A.D. 400-1450) in Europe, rigid fans were used. Metal disks on long handles shooed flies during church ceremonies, and the rich used ornamental fans fashioned from parchment mounted on ivory, gold, or silver handles.

Beginning in the fifteenth century, Portuguese traders brought large quantities of folding fans from Asia, and by the seventeenth century they were highly popular. Although the fans exported to Europe were of much poorer quality than those used in Asian countries, they were much admired by their European purchasers. During the reign of King Louis XV (1710-1774) in France, even men carried dainty fans.

Fans were decorated in accordance with the styles of the day and ranged from simple to ornate, some bearing reproductions of famous paintings. Their sizes ranged from the 8-inch version popular in the early 1800s to as large as 20 inches during the Victorian era (1837-1901). The most expensive fans were made from parchment or silk, with handles of carved ivory, tortoise shell, horn, bone, or sandalwood.

Electricity Powers Modern Fans

The use of handheld fans for cooling purposes and as decorative accessories died out, for the most part, after the nineteenth century. The electric fan was first produced commercially (in a two-bladed desk version) by Schuyler Skaats Wheeler for the Crocker & Curtis Electric Motor Company in 1882. It proved highly effective and less physically taxing as a cooling aid.

Over the next century electric fans were improved: floor and window models appeared, oscillating (turning) cases sent cool air over a larger area, and coatings quieted blade noise. Today, most inexpensive household fans have blades composed entirely of **plastic**. Aside from being cheaper to produce, they are not as likely to injure fingers accidentally inserted in the path of the blade.

Fats and oils

Fats and oils are energy-rich compounds that are basic components of the normal diet. Both have essentially the same chemical structure—a mixture of fatty acids combined with glycerol (a trihydroxy alcohol)—and are insoluble (do not dissolve) in water. However, while fats remain solid (or semisolid) at room temperature, most oils very quickly become liquid at increased temperatures. Animal fats and oils include butter, lard, tallow, and fish oil. Plants provide a number of oils, such as cottonseed, peanut, and corn oils.

Saturated fats usually come from animal sources, have very high melting points, and can often increase the body's blood cholesterol. By doing so, saturated fats are thought to contribute to the onset of certain diseases such as atherosclerosis (heart disease).

Fats have two main functions: they provide some of the raw material for synthesizing (creating) and repairing **tissues** and they serve as a concentrated source of fuel energy. Fats, in fact, provide humans with roughly twice the energy, per unit weight, as do carbohydrates and proteins. Fats are not only an important source of day-to-day energy, they can, if not immediately needed, be stored indefinitely as adipose (fat) tissue in case of future need.

Fats also help in other ways. They transport fat-soluble **vitamins** throughout the system. They cushion and form protective pads around delicate organs, such as the heart, liver and kidneys. They make up the layer of fat under the skin that helps insulate the body against too much heat loss. They even add to the flavor of other foods that might otherwise be inedible.

But if normal amounts of fat in the diet are essential to good health, unnecessarily high amounts can lead to a number of problems. For instance, a certain amount of excess adipose tissue can be valuable during periods of illness, overactivity, or food shortages. However, too much adipose tissue can not only be unsightly but can overwork the heart and put added stress on other parts of the body. High levels of certain circulating fats may not only lead to atherosclerosis, but have been linked to other illnesses, including cancer.

How much fat in the diet is considered too much? In the past, nutritionists considered reasonable a diet that obtained 40 percent of its calories from fats. Today, however, they recommend that no more than 30 percent (and preferably even less) come from fat. In healthy adults, too, body fats typically should make up no more than 18 to 25 percent of the body weight in females and 15 to 20 percent in males.

See also **Molecular structure**

Fat substitute

Although fat contributes 35 to 40 percent of the daily calorie intake of Americans, most physicians agree that 30 percent would be healthier. As people realized the need to cut fat intake, food manufacturers saw the developing possibility of a large, new market: fat substitutes.

In the 1980s, the NutraSweet Company (owned by Monsanto and based in Deerfield, Illinois) was especially interested in fat substitutes because it needed a new high-profit product. The patents on its main moneymaker, the **artificial sweetener** NutraSweet, would start expiring in 1992.

Some companies developed starch and water mixtures to replace more than half the fat in products such as salad dressings. Hellman's Light mayonnaise used this emulsified (liquid) starch.

As people realized the need to cut fat intake, food manufacturers saw the developing possibility of a large, new market: fat substitutes.

Then a food scientist at NutraSweet, Norman Singer, developed an emulsified protein substance. Singer mixed and heated proteins from milk and egg whites, and then made the mixture into round microscopic particles. As they roll on the tongue, the particles create the creamy feel of fat—but with only 15 percent of the calories of fat. In February 1990 NutraSweet's product, called Simplesse, became the first fat substitute approved by the Food and Drug Administration (FDA) for use in frozen desserts. NutraSweet immediately launched its Simplesse-based reduced-calorie Simple Pleasures frozen dessert. While Simplesse works well for some products, it has a large drawback: it cannot be used for cooking because heat causes it to stiffen.

A potential solution is Procter & Gamble's olestra. Unlike Simplesse, which is made from natural ingredients, olestra is a synthetic (artificial) compound of sugar and fatty acids. It has no cholesterol or calories, and is able to pass through the body without being absorbed. Olestra can be used in cooked foods. Development of olestra took more than 20 years, and Procter & Gamble applied for FDA approval of the imitation fat in 1987. Because olestra is a synthetic food additive, Procter & Gamble's application triggered lengthy safety studies.

Many other companies are developing fat substitutes, some with gelatin-water mixtures, and are currently test-marketing them in butter and margarine products. More artificial fats are certain to appear on the market in the near future.

Fax machine

The facsimile, or fax, machine is both a transmitting and receiving device that "reads" text (printed words), maps, **photographs**, fingerprints, and drawings. The fax communicates via telephone line. Since the 1980s, fax machines have undergone rapid development and refinement and are now indispensable communication aids for newspapers, businesses, government agencies, and individuals.

Since the 1980s, fax machines have undergone rapid development and refinement and are now indispensable communication aids.

The use of faxes, and fax technology itself, remained limited until the mid-1980s. Until then fax machines needed special heat-sensitive paper. Updated models from the 1990s use plain paper, which does not curl and gives better quality. Another improvement is the invention of a scrambler, an encoder that allows the sender to make sure his or her documents go only to the intended receiver. This feature is important when the documents are about highly sensitive government projects or secret industrial or business dealings.

How the Fax Works

Some fax machines are part of telephone units while others stand alone. Still others are part of personal computers. These last models contain a fax board, an electronic circuit that allows the computer to receive messages.

In the most common models, the user inserts the material to be transmitted into a slot, then makes a telephone connection with another facsimile machine. When the number is answered, the two machines make an electronic connection. A rotating drum pushes the original document past an optical scanner. The scanner reads the original document either in horizontal rows or vertical columns and translates the printed image into a pattern of several million tiny electronic signals, or pixels, per page. The facsimile machine can adjust the number of pixels so that the sender can control the sharpness and quality of the transmission.

Within seconds, the encoded pattern is converted into electric current by a photoelectric cell, then travels via telegraph or telephone wires. The receiving fax, which matches the sender's signal, produces an exact replica of the original.

See also **Computer, digital**

Fermentation

In the absence of the gas **oxygen**, certain living things are capable of breaking down **carbohydrates** (starches and sugars) to form alcohol and **carbon dioxide** gas. This process is known as anaerobic respiration or fermentation. Fermentation has been used for centuries in the production of certain foods and beverages.

The fermentation process is caused by **enzymes**, tiny catalysts that create chemical reactions similar to the digestive enzymes in the human body. Certain enzymes act on starch to break it down into smaller units of sugar. Then other enzymes convert one kind of sugar molecule to another. Still other enzyme reactions break the sugar molecule into ethyl alcohol and carbon dioxide gas.

Yeast is traditionally added to liquids derived from grains and fruits to brew beer and wine. The natural starches and sugars provide food for the yeast, and during fermentation the desired alcohol is released.

In Food

The carbon dioxide and alcohol by-products of fermentation have been used in human enterprise for centuries. The **yeast** *Saccharomyces cerevisiae* is traditionally added to liquids derived from grains and fruits to brew beer and wine. The natural starches and sugars provide food for the yeast and during fermentation the desired alcohol is released. In China, for thousands of years traditional soy sauce or *shoyu* was brewed by adding the fungus *Aspergillus oryzae* to a mixture of boiled soybeans and wheat and allowing it to ferment for about a year.

As Energy

In recent times, yeasts have been used to create alternative energy sources. Yeasts are placed in large fermentation vats containing organic material (things that were once alive). During fermentation the yeast changes the organic material into ethanol fuel. Researchers are working on developing yeast strains that will convert even larger organic biomasses into ethanol more efficiently.

The fermentation process is not limited to microorganisms. It takes place on occasion in animals. In human beings, fermentation occurs in muscle tissue during periods of exercise. When energy requirements are high during strenuous exercise, oxygen cannot be delivered fast enough. To continue supplying the muscle with energy, sugar molecules in the **tissue** are converted to lactic acid. However, the buildup of lactic acid causes fatigue and, even after exercise ceases, breathing continues to be heavy to get more oxygen back into the tissue.

Ferris wheel

A Ferris wheel is a carnival ride consisting of a large, vertical wheel rotating around an axle. The ride was invented by engineer George Washington Gale Ferris for the World's Columbian Exposition in Chicago, Illinois, in 1893.

Ferris was born February 14, 1859, in Galesburg, Illinois. As a child he lived in Carson City, Nevada. He graduated from Rensselaer Polytechnic Institute in Troy, New York, in 1881 and set to work designing railroads, **bridges**, and **tunnels.** In 1885 he began inspecting structural steel for the Kentucky and Indiana Bridge Company. Structural steel is used in building supports. He took what he learned on the job and started his own business, G. W. G. Ferris & Company, which acted as a consultant to steel users.

In 1892 Daniel Burnham (1846-1912), planner for the World's Columbian Exposition in Chicago, announced he was looking for a display that would rival the Eiffel Tower built for the Paris Exposition three years earlier. Ferris reacted by producing a design for a giant upright wheel. It was both an engineering and an aesthetic marvel.

Invented for the 1893 Columbian Exposition, the first ferris wheel was meant to rival the Eiffel Tower, which had been unveiled three years earlier at the Paris Exposition.

The original Ferris wheel towered 250 feet over the exposition and could hold 60 people in each of its 36 cars. It weighed 2,100 tons, had a forged steel axle 33 inches by 45 feet, could carry 150 tons of riders, and was driven by two 1,000-horsepower reversible **steam engines**. The combination of its lights and the rhythm of its whirling motion was similar to the visual impact of a steam-driven paddle wheel.

To finance the venture, Ferris raised $250,000 through a stock issue. His wheel made a profit of more than $1 million. Ferris died three years later, in 1896.

See also **Train and railroad**

Fertilization

Fertilization occurs when the (male) sperm and the (female) egg, or ovum, are brought together. In many aquatic (sea) and amphibian species, the eggs are released by the female, while the male deposits his sperm on or near them. However, in most mammalian species (including humans) the male uses a special organ to deposit the sperm safely within the female. The sperm then swims randomly until it dies or encounters an egg.

Once the sperm and egg meet, the fertilization process can begin. As the tip of the sperm touches the outer surface of the ovum, certain chemical reactions are set into motion. The two bodies are fused together, while the contents of the sperm are allowed to pass into the egg. There the sperm mingles with the ovum. This process is called activation. Shortly after, the surface of the egg changes to prevent other sperm from entering, thus insuring fertilization by one sperm only.

Once the egg is "activated" it begins to divide—first into two cells, then four, then eight, and so on. These multiplying cells will eventually form the fetus. The rapidly developing egg is now called a zygote, marking the end of the fertilization process.

Fertilizer, synthetic

Fertilizers replace the minerals absorbed from the soil as plants grow. These minerals include **phosphorus**, **potassium**, **nitrogen**, sulfur, calcium,

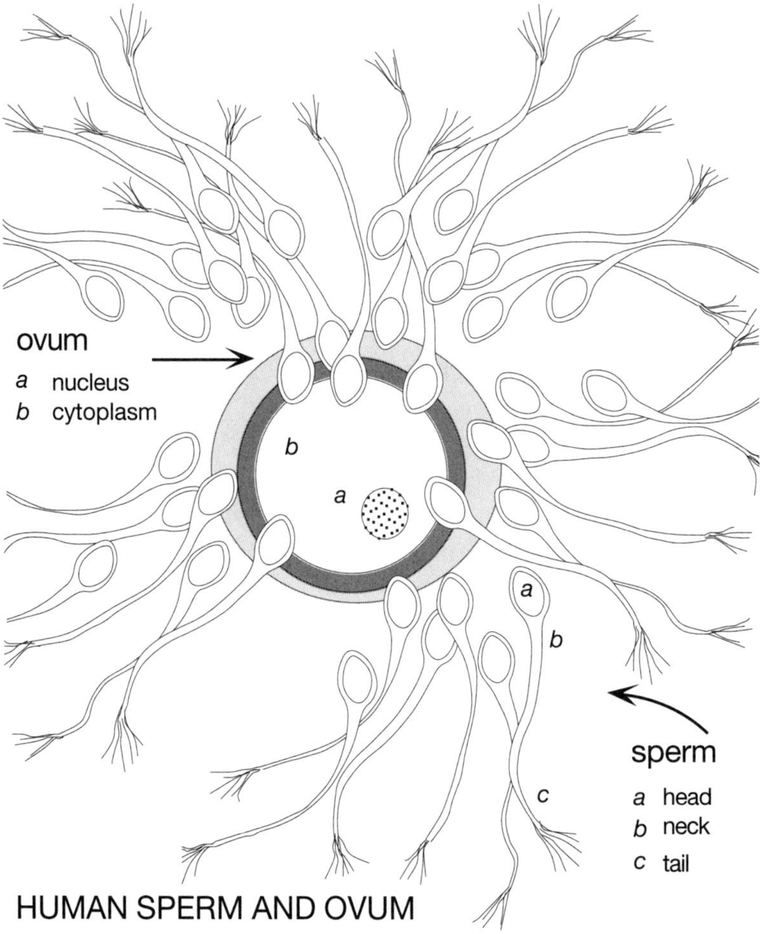

See ***Fertilization*** *entry on page 453. Fertilization occurs when the (male) sperm and the (female) egg, or ovum, are brought together.*

iron, and magnesium. Trace elements or elements that plants use in smaller amounts include copper, boron, zinc, manganese, and cobalt. Plants also require carbon, hydrogen, and oxygen, which are usually supplied through the air and water.

Depending on the plant and the soil, some or all of these elements may need to be added. If the minerals are not replenished, the soil will not support the crop growth. This was not such a problem in the United States or Canada during the early years of westward expansion because farmers would simply move to more fertile land when they had depleted the soil's mineral supply. However, replenishing these minerals was critical in Britain, Europe, and Asia, where farm land was in shorter supply. So farmers and the agricultural industry turned to chemists for help in making artificial fertilizers.

Since exhausted soil most commonly needs potassium, phosphorus, and nitrogen, these are the nutrients that chemists set out to replace.

Potassium

Natural reserves of potassium are ample. Originally the largest supplies were in Germany, but today potassium chloride mines in New Mexico and California supply much of the potassium used in synthetic (artificial) fertilizers. Sulfur is present in the phosphate or ammonium fertilizers, usually as an impurity of the production process rather than as a deliberate additive. Calcium is currently supplied by dolomite, limestone, burned lime, or wood ash. Iron and magnesium are also added to fertilizers, as are the trace elements.

Phosphorus

Phosphorus is also needed in large quantities for plant growth. Today most phosphorus is supplied from rocks containing superphosphates or triple superphosphates. These are produced by applying sulfuric acid to natural phosphorus found in rocks.

Agricultural spraying. Farmers and the agricultural industry turned to chemists for help in making artificial fertilizers to augment (increase) depleted mineral supplies in soil.

Nitrogen

While fertilizers containing phosphorus were being produced, nitrogen-containing fertilizers were slower in developing. Not until the early 1900s did a German chemist, Fritz Haber, develop the technique for producing ammonia (which contains nitrogen) synthetically. He combined **hydrogen** with nitrogen and a catalyst (something that sparks a reaction) at a high temperature and pressure to form synthetic nitrogen. By 1913 Carl Bosch, a German chemist working for BASF, developed the industrial ammonia production process. The first ammonia factory produced 9,000 tons of ammonia in its first year. The production techniques developed for the fertilizer were later applied to the petroleum industry.

Fiberglass

During World War I, the Germans began manufacturing fiberglass as a substitute for asbestos.

Fiberglass is an artificial material used in everything from home insulation to automobile repair. Fiberglass consists of very fine threads of glass, sometimes combined with other materials, loosely bunched together in woolly masses. Fiberglass resists burning, and will not decay, stretch, or fade. It is also an excellent insulator, as the tiny spaces between the fibers trap air, preventing the flow of heat. Such properties make it desirable for weaving into cloth for curtains and tablecloths. Since it is flexible and strong, fiberglass is ideal for combining with **plastics** for **automobiles**, boat bodies, and fishing rods. Fiberglass is also used for packing into woolly bulk form for air filters.

Ancient Egyptians wove glass fibers into their vases and containers, creating a decorative trim by winding the glass around a core of glazed clay. But fiberglass manufacture on a large scale did not take place until the twentieth century. During World War I (1914-18), the Germans faced a shortage of asbestos. They began manufacturing fiberglass as an excellent substitute insulator. In the United States in the early 1930s, the Owens Illinois Glass Company and the Corning Glass Works conducted experiments that resulted in the successful commercial manufacture of fiberglass.

Fiber optics

Since the late 1950s, optical fibers have emerged as revolutionary tools in the fields of medicine and telecommunications. These fibers can transmit light pulses containing data up to 13,000 miles (20,917 km)—and do so without significant distortion. The fibers also permit the "piping" of light into the body, allowing doctors to see and diagnose conditions without the use of surgery.

Optical fibers operate by continuously reflecting light (and images) down the length of the glass core.

Making Fiber Optic Cables

The manufacturing of optical fibers consists of coating the inner wall of a silica glass tube with 100 or more successive layers of thin glass. The tube is then heated to 3,632° F (2,000° C) and stretched into a strand of thin, flexible fiber. The result is a clad fiber, approximately 0.0005 inch (0.0127mm) in diameter. By comparison, a human hair measures 0.002 inch (0.0508 mm).

Using Fiber Optics

Fiber optics were first used in medicine. In the late 1950s, Narinder S. Kapany hit upon the idea of building an **endoscope** that could see around the twists and turns in a patient's body by using fiber optic bundles. His fiberscope consists of two bundles of fiber. One carries light down to the site to be studied. The other bundle carries a color image of the site

Optical fibers have emerged as revolutionary tools in the fields of medicine and telecommunications.

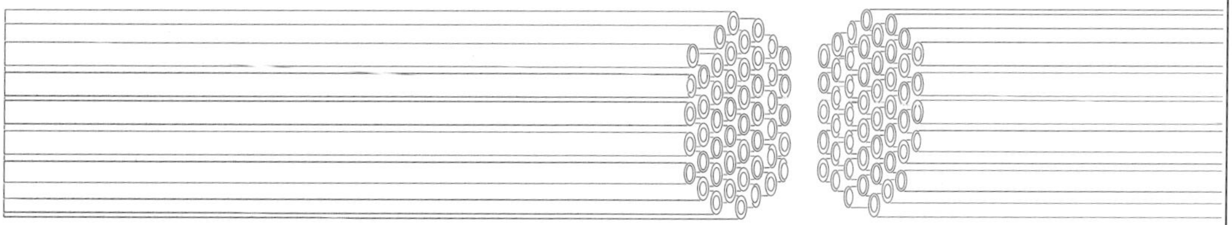

Pulses of light transmit data through transparent fibers. Optical fibers are used in medicine to view internal organs.

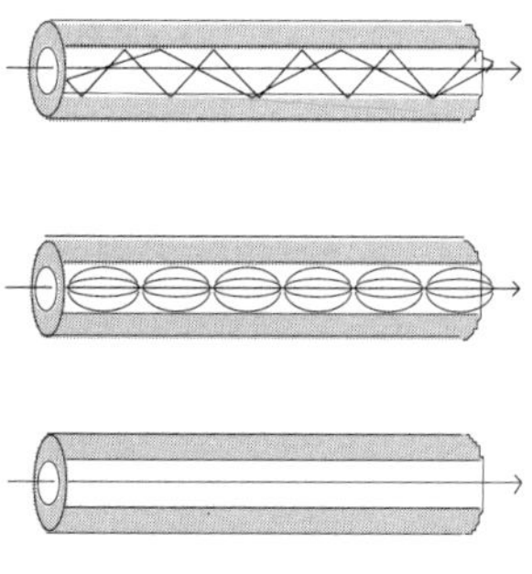

Types of lightwaves traveling through fibers.

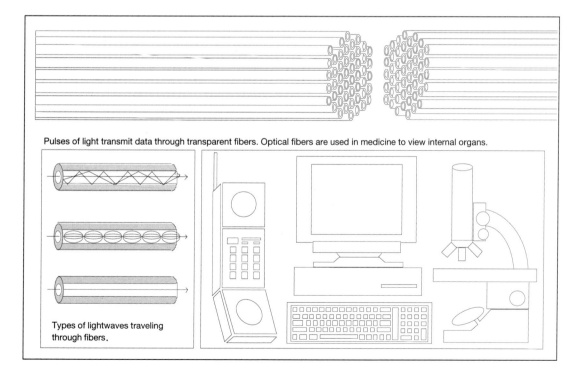

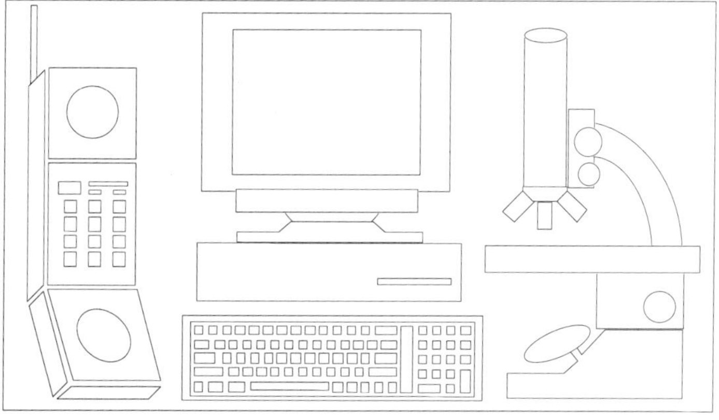

back to the physician. Because of its small size and flexibility, the fiberscope can be used to view many areas inside the body, such as the heart, vein, artery, and digestive systems.

Fiber optics have changed forever the way we communicate and the way we practice medicine.

Optical fibers were first used in the field of telecommunications in 1966. Today a telephone conversation can be carried over optical fibers by a method called digital transmission. This is achieved by first converting sound waves into electrical signals, each of which are then assigned a digital code of 1 or 0. The light carries the digitally encoded information by emitting a series of pulses: a 1 would be represented by a light pulse, while a 0 would be represented by the absence of a pulse. At the receiving end, the light waves are converted back into electronic data, which are then converted back into sound waves.

By using digital transmission, telecommunications systems carry more information farther, over a smaller cable system than its copper wire predecessor. A typical copper bundle measuring 3 in. (7.62 cm) in diameter can

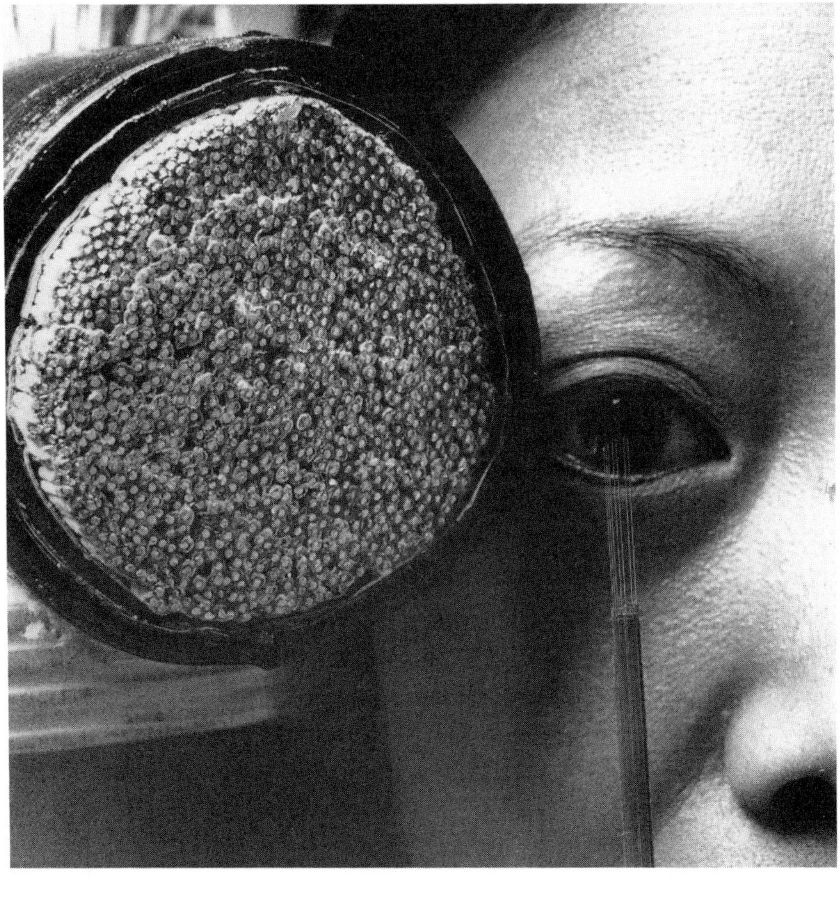

A one-quarter-inch-wide optical fiber can carry the same amount of data as this copper telephone trunk cable measuring 3 inches in diameter.

be replaced by a 0.25 in. (0.635 cm) wide optical fiber carrying the same amount of data. This improvement becomes important in areas where telephone cables must be placed underground where space is limited.

The tiny size of optical fibers also allows for a significant reduction in the weight of a particular system. The reduced weight is beneficial in systems that require rapid deployment of information, such as in military communications and in **aircraft** instrument wiring. Replacing the copper wiring on a jet aircraft can save up to 1,000 pounds (454 kg), allowing for more economical fuel consumption. Optical fibers are also immune to electromagnetic interference, making them roughly 100 times more accurate than copper. They typically allow only one error in 100 million bits of data transmitted.

Optical fibers have proven to be an ideal method of transmitting high-definition **television** (HDTV) signals. Because its transmissions contain twice as much information as those of conventional television, HDTV allows for much greater clarity and definition in its picture. However, standard transmission technology cannot transmit so much information at once. Using optical fibers, the HDTV signal can be transmitted as a digital light-pulse, providing a near-flawless image. The HDTV reproduction is far superior to broadcast transmission, just as music from a digital CD is superior to that broadcast over FM **radio**.

Fiber, synthetic

Synthetic fibers are strands made of **polymers**, a type of plastic. After being created chemically, synthetic fibers are washed, dried, dyed, and woven. Uses for such fibers range from nylon stockings and clothing to cables and tire reinforcement.

Uses for synthetic fibers range from nylon stockings and clothing to cables and tire reinforcement.

The first patent for synthetic fiber was granted to George Audemars in 1855. A related patent was granted to Sir Joseph Swan in 1880. Both of these men produced fibers—though not very strong ones—from cellulose (a carbohydrate). A later, stronger version was called rayon. Many other synthetic materials are in use today, including:

- Acrilan is an acrylic fiber used in fabrics and may be blended with wool or cotton to form clothing, carpeting, linens, draperies, and upholstery. Fabrics made from acrilan resist mildew, moths, and wrinkling. They also tend to dry quickly.
- Acrylonitrile fabric, which resembles soft wool, is used for sails, cords, blankets, and clothing.

- **Dacron** is a wrinkle-resistant textile used in the clothing industry.
- **Kevlar**, an offshoot of the U.S. space program, is incredibly strong and light weight. Commercial uses today include parachutes and kite string.
- **Nylon** was developed by Du Pont Company researchers as a substitute for silk. Nylon is used in clothing, laces, toothbrushes, sails, fish nets, and carpets.
- Orlon can be woven or knitted, usually into bulky garments, and is used in upholstery and carpets.
- **Vinyl** was the first plastic fiber produced on a large scale in the United States, and was patented in 1947.

See also **Plastic; Toothbrush and toothpaste**

Fingerprinting

Each of the 5.5 billion people on Earth possess unique fingerprints.

A fingerprint is made up of the pattern of ridges on a person's fingertips. Each fingerprint is unique and its pattern never changes. These facts make fingerprinting a foolproof method of identification. By studying the number and sequence of the ridges in the fingerprint patterns, fingerprint specialists can positively match an individual to a set of prints.

With more than 200 million fingerprints on record at the Federal Bureau of Investigation (FBI), matching fingerprints is not an easy job. Categorizing of fingerprint patterns makes the matching process easier. There are three general types: whorl, arch, and loop. Eight subcategories have also been developed to define the different combinations these general patterns can create.

In Police Work

Fingerprints are especially helpful in detecting crimes. Criminals can change their names and their appearance but not their fingerprints. Many have attempted to do so with everything from sandpaper to strong acids! Their best efforts have produced temporary pain or permanent injury.

Detectives look for three types of prints at a crime scene: visible, molded, or latent. Visible fingerprints are left by fingertips coated with paint, grease, or some other visible substance. Molded fingerprints are left

Fingerprints in History

Establishing exactly how long fingerprinting has been around is difficult. Prehistoric carvings resembling fingerprint patterns have been found in cliffs, caves, and even ancient clay tablets. The Chinese began to use thumbprints to sign documents long before the birth of Christ, but probably as a legal signature rather than a means of identification.

In 1823 J. E. Purkinje (1787-1869), an anatomy professor, published a paper noting the diverse ridge patterns on human fingertips, but did not suggest his observations as a way to identify people. In 1858 Sir William J. Herschel devised a workable fingerprint identification system thought to be the first of its kind. In the 1880s, Francis Galton obtained the first extensive collection of fingerprints for his studies on heredity. He also established a bureau for the registration of civilians by means of fingerprints and measurements.

In 1891 Juan Vucetich, an Argentine police officer, was believed to have created the first usable fingerprint identification system that could be applied to criminal investigation. And a few years later Sir Edward R. Henry of Great Britain developed a more simplified fingerprint classification system still used today.

in soft items such as wax or putty. Latent or hidden prints are left by the skin's natural oils and secretions. These may not be visible to the naked eye but can be brought out by special powders and chemicals.

In addition to their use in criminal investigations, fingerprints are useful in other areas. For example, they can be used to identify accident victims, armed forces personnel, government employees, and amnesia (loss of memory) victims.

Taking Fingerprints

Fingerprints are recorded by rolling the fingertips on an ink-covered surface. Then each finger is pressed onto a standard card divided into ten squares, one print per square. As a check for the proper sequence, all five fingers are simultaneously pressed onto another large square.

The current fingerprinting system far surpasses identification methods that preceded it. Before fingerprinting was introduced, people had to be identified by tattoos, brands, **photography**, body measurements, and other inefficient means that resulted in many misidentifications.

Fire extinguisher

George William Manby (1765-1854) invented the first fire extinguisher in 1813. Manby, who had been a member of the British militia (army), was inspired after seeing Edinburgh, Scotland, firemen trying to reach the upper floors of burning buildings.

Manby's extinguisher consisted of a four-gallon copper cylinder that held three gallons of water. The remainder of the space in the cylinder was filled with compressed air. When used, the compressed water would be forced out through a tube in the cylinder and could be pointed at the fire.

Modern fire extinguishers rely on carbon dioxide to force out the extinguishing agent (water, foam, or powder).

In 1866 Frenchman François Carlier invented an extinguisher that used a chemical reaction to force the water out. A Russian, Alexander Laurent, developed an extinguisher with a chemical reaction meant to be used on oil-based and electrical fires. During World War II (1939-45), a commercial product called Aero Foam was developed from soy protein and used by the United States military.

Modern fire extinguishers rely on carbon dioxide to force out the extinguishing agent (water, foam, or powder).

Entire buildings are now required to have fire extinguishing systems. These consist of a network of sprinklers installed in the ceilings, and are heat-activated.

See also **Firefighting equipment**

Firefighting equipment

Firefighters

Among the earliest attempts to organize firefighting was the ancient Egyptians' gathering of volunteer firefighters. The Romans used slaves stationed in strategic locations to spot and douse (put out) fires. The Greeks and Romans even devised primitive fire engines—small human-powered water pumps mounted on wheels or skids. But for many centuries firefighting essentially consisted of little more than bucket brigades, where people stood in a line between the town well and the blaze, passing filled buckets up one side of the line and empty ones down the other.

Water Wagons

The great London fire of 1666, which destroyed nearly 13,000 buildings, dramatized the need for a way to deal with fire emergencies. Hand-operated pumps on wheels drawn by humans again came into use, and were replaced in the 1800s by horse-drawn wagons. These early pumps could produce streams of water of no more than 50 feet (15.25 m). In 1830 John Giraud of Baltimore invented a pump with a chamber of compressed air that ultimately was capable of boosting the stream of water delivered through the leather hoses to about 200 feet (62 m).

Alarms

The first fire alarms were invented by Ithiel Richardson in 1830. He

Fighting a fire in New York City, 1993. Modern fire trucks are fully equipped with life-support systems and a range of tools for battling fires.

The Pilgrims, and later the pioneers, relied on bucket brigades to put out fires.

ran strings through rooms of a building. When the strings burned through, they set off a central bell. Rufus Porter used a metal bar that would drop onto a bell apparatus when heated. Welshman George William Manby invented the first portable **fire extinguisher** in 1813. The four-gallon copper vessel contained water and compressed air. Later extinguisher designs were chemically activated and could put out electrical and oil fires.

Steam Is Introduced

A rescue ladder attached to a fire wagon on a revolving base was first introduced in 1840. That same year P. R. Hodge developed the first version of a steam-powered fire engine. However, it was not until 1852 that A. B. Little's improved steam-powered engine was accepted for use by firefighters.

Modern Fire Trucks

Gasoline-powered engines came into use in the 1900s. No longer was it necessary to "fire up" the engines to build up steam before responding to the fire. The entire range of mechanical equipment could be operated automatically by electricity. Modern fire trucks are fully equipped with life-support systems and a range of tools for battling fires, including pumps capable of throwing 750 to 2,000 gallons (2,838.75 to 7,570 l) of water per minute, steel water towers, and ladders that can extend over 100 feet (30.5 m) into the air. Sophisticated alarm systems are activated directly from business and industrial sites. Many cities depend on the computerized Enhanced 9-1-1 Telephone System, which allows individuals quick-dial access to emergency services.

Special Cases

Different measures are needed for other types of fires. The main course of action in fighting a forest fire, for example, is to contain (control) it rather than extinguish it. Firefighters are flown or driven in to create a fire line. The line is formed by cutting, dousing (wetting), or burning a break in the trees in the hope of stopping the fire's advance.

Ships at sea must have their own firefighting capabilities. In harbors, fireboats can be called into action.

See also **Fire extinguisher; Internal combustion engine**

Fireworks

The first fireworks were most likely created in China during the tenth century and used during ceremonies. They depended on black powder (gunpowder) for their vivid displays. The first firework was a projectile, resembling the modern-day Roman candle and spewing balls of fire from a bamboo tube.

Although the Chinese were the originators of this device, the Arabs hold claim to developing, in 1353, the first gun. It was a bamboo tube reinforced with iron to withstand the explosive pressure of the compacted powder. However, it was an Englishman, Roger Bacon, who wrote the first specific instructions for the preparation of black powder in 1242. Bacon's recipe has since been altered and improved upon for both military and industrial uses.

Entertainment Value

Perhaps the first person to ardently promote and cultivate the stunning visual and dramatic possibilities of fireworks was King Louis XIV of

Fireworks shot off over New York City in June 1991. The best fireworks displays emphasize pageantry, ornateness, and surprise.

France, the Sun King who reigned from 1643 to 1715 and whose palace at Versailles formed the perfect backdrop for his lavish, fireworks-punctuated galas. In much the same baroque spirit today, the best fireworks displays emphasize pageantry, ornateness, and surprise, all with the assistance of computerized choreography, split-second electrical firing, elaborate one-of-a-kind set pieces, and massive lines of mortars from which the large shells (stars, chrysanthemums, comets, peonies, salutes, etc.) are launched.

The history of fireworks, a peaceful invention, is intertwined with that of early instruments of war.

The six basic fireworks colors (along with their key ingredients) are white (magnesium or aluminum), yellow (sodium salts), red (strontium nitrate or carbonate), green (barium nitrate or chlorate), blue (copper salts), and orange (charcoal or iron). Fireworks are widely restricted or prohibited for private use. They range in size from the 3/4 inch-long (2 cm) ladyfinger firecracker to the world record Fat Man II, a 720-pound (326 kg), 40.5 inch-diameter (102 cm) shell fired near Titusville, Florida, on October 22, 1977.

The Technical Side

All fireworks consist of a fuel source (such as gunpowder), an oxidizer (oxygen source), a fuse, and color-producing compounds. The gunpowder itself, a ground-up mixture of potassium nitrate, sulfur, and charcoal, is the most important variable and determines the speed, height, and bursting power of the charge. The large-scale manufacture of fireworks is carried on today in a number of countries, including Japan, France, England, Spain, Italy, and the United States. Major fireworks displays are enjoyed the world over and have become irreplaceable centerpieces of such festivals, observances, and holidays as the Fourth of July, New Year's Eve, Mardi Gras, and the Chinese New Year.

For fireworks manufacturers—almost all of which are longstanding, family owned businesses—black powder recipes are highly prized and carefully guarded trade secrets.

Flamethrower

Ancient people were well acquainted with the use of fire as a weapon. Over hundreds of years they launched burning arrows, dumped flaming oil, and tarred criminals with hot pitch (a dark, sticky substance made from wood, coal, or petroleum).

By the twentieth century, the use of fire had grown more sophisticated. The Germans first used a flamethrower during World War I (1914-

18) against the French. The German weapon consisted of a pack containing a liquid agent worn by the soldier, a hose connected to the liquid agent, and a nozzle that measured the amount of liquid released. The liquid was squirted out of the hose and ignited by an attached spark device.

The British army in World War II (1939-45) carried an effective flamethrower that was capable of burning continuously for ten seconds. This device could shoot unlit fuel onto a target, which could be ignited later. The British also developed an **armored vehicle**, the Crocodile, which had flamethrowing ability.

Flashlight

Now a common household item, the lowly flashlight was once considered a toy. When it was introduced in 1898 at an electrical show in New York, the flashlight weighed more than six pounds, and its battery alone was half a foot long. Although patents for the device were issued to American Electrical Novelty and Manufacturing Co. in the 1890s, no single person has laid claim to its invention. The inventor of the toy electric train, American Joshua Cowen (1880-1965), built an early flashlight, but used it merely to shine on potted flowers.

The lowly flashlight was once considered a toy.

Today's lightweight, powerful flashlights are considerably more convenient and useful, thanks to improvements in light bulbs, batteries, and controls. In most flashlights, which use incandescent (glowing) electric light bulbs, the light is focused into a narrow beam by a reflector and a lens. Small fluorescent models are also available, and some flashlights use extremely brilliant arc lamps that can illuminate objects in darkness half a mile (0.8 km) away.

See also **Battery, electric**

Fluorescence and phosphorescence

Substances that glow in the dark, called "phosphorescent" (fos-for-es-ent) materials, have many practical applications today. Clocks and watches, for example, often have their numbers and hands coated with phosphorescent paints so we can see what time it is in the dark. Emergency doors and stairways are also highlighted with these paints so that people can find their way out in case of a power failure.

Phosphorescent paint coats the hands and numbers of glow-in-the-dark watches.

"Fluorescence" describes the visible glow produced by materials exposed to ultraviolet rays, an invisible form of light. This phenomenon is closely related to phosphorescence since both are caused when a chemical's atoms become "excited" by **ultraviolet radiation**. The main difference is that fluorescent chemicals lose their glow almost immediately, whereas phosphorescent materials can continue to glow for several hours after their atoms are excited.

Modern fluorescent lighting is created by coating the inside of a glass tube with phosphors, which convert the lamp's ultraviolet rays to visible light. Fluorescent chemicals are also used in common household products such as eye drops and tooth brighteners. Scientists can identify compounds by their characteristic fluorescent "fingerprint"—the wavelength at which the substance begins to glow.

Fluoride treatment, dental

Fluoride in the public water system has cut the number of cavities in Americans.

Fluoride is a chemical found in many substances. In the human body, fluoride acts to prevent tooth decay by strengthening tooth enamel and inhibiting the growth of plaque-forming bacteria. After researchers discovered this characteristic of fluoride, fluoridation—the process of adding fluoride to public water supplies—began.

It all started with Frederick S. McKay, a Colorado Springs, Colorado, dentist, in the early 1900s. McKay noticed that many of his patients had brown stains, called "mottled enamel," on their teeth. McKay set out to find the cause. By 1916 McKay and a team of other researchers believed the mottling was caused by something in the patients' drinking water. By 1928 the team concluded that mottling was linked to reduced tooth decay.

In 1931 the link was made between mottling and high concentrations of fluoride in the water. Experiments showed that the ideal level of fluoride was one part per million. This was enough to stop decay but too little to cause mottling.

Fluoride in Water

The U.S. Public Health Service grew interested in fluoride and, following safety tests on animals, it conducted field tests. In 1945 the public water systems of Newburgh, New York, and Grand Rapids, Michigan, became the first ever to be artificially fluoridated with sodium fluoride. Simultaneously, a group of Wisconsin dentists led by John G. Frisch inaugurated fluoridation in their state.

Results of these tests seemed to show that fluoridation reduced dental cavities by as much as two thirds. Based on those results, the United States Public Health Service recommended in 1950 that all United States communities with public water systems fluoridate. Later that year the American Dental Association (ADA) followed suit, and the American Medical Association added its endorsement in 1951.

Fluoride is toxic, or poisonous, in large amounts.

Objections to Fluoridation

Even though almost the entire dental, medical, and public health establishment favored fluoridation, the recommendation was immediately controversial, and has remained so. Opponents objected to fluoridation because of possible health risks (fluoride is toxic, or poisonous, in large amounts). They also objected to being deprived of the choice whether or not to consume a chemical. Despite the opposition, nearly 60 percent of people in the United States now drink fluoridated water. Fluoridation is also practiced in about 30 other countries.

Fluoride in Toothpaste

The initial claims that fluoridation of drinking water produced two-thirds less tooth decay have been modified to about 20 to 25 percent reduction. Other ways of applying fluoride have been developed. In the 1950s Procter & Gamble (P&G) had the idea of adding the chemical to toothpaste. Procter & Gamble's new "Crest—with Fluoristan" was launched in 1956 with an advertising blitz that included the popular line "Look, Mom—no cavities!" Four years later, P&G celebrated when the Council on Dental Therapeutics of the American Dental Association (ADA) gave Crest its seal of approval as "an effective decay-preventive dentifrice." The ADA now estimates that brushing with fluoride-containing toothpaste reduces tooth decay by as much as 20 or 30 percent.

In addition to toothpaste, fluoride can be taken in tablet form, and as a solution either "painted" directly onto the teeth or swished around as a mouthwash.

See also **Toothbrush and toothpaste**

Fluorine

Fluorine is a pale yellow, corrosive gas. Its atomic symbol is F and its atomic number is 9. Fluorine is a ferociously reactive element that combines, often violently, with all other elements except some of the inert (unreactive) gases. Even water will burn with a bright flame in an atmosphere of fluorine gas.

A form of fluorine is used in fluoride toothpaste.

Many of fluorine's compounds are toxic (poisonous) and can cause deep, severe burns on contact. Yet fluorine exists in many harmless compounds as well. Most of it occurs in the mineral fluorspar, which is found throughout the world.

In the United States, Illinois produces more than half the nation's fluorspar. Fluorine is also a constituent (part) of the mineral cryolite, which is found in commercial quantities only in Greenland.

Fluorine possesses the smallest and lightest atoms of the halogen family, a group of elements that readily combines with metals to form salts. The element reacts with most inorganic compounds to form fluorides such as **uranium** hexafluoride. This compound, which is used to prepare uranium for atomic bombs and nuclear reactors, became important during World War II (1939-45) and spawned the commercial fluorine production industry.

Since the turn of the twentieth century, scientists have developed techniques for handling fluorine safely and transporting it as a liquid. Since World War II, fluorine and its compounds have found many industrial applications. For example, hydrofluoric acid is used to etch glass for products such as light bulbs. The oil industry uses hydrogen fluoride as a gasoline additive.

Fluorine is also a constituent of **polymers** (plastics) such as Teflon that are used to coat frying pans and other products. And the chemical industry uses crystals of calcium fluoride to analyze compounds by bending and focusing infrared light onto them.

Fluorine-containing compounds called **CFCs (chlorofluorocarbons)** have traditionally been used in refrigerators, air conditioners, and aerosol sprays. However, these compounds have been linked to the depletion of Earth's protective **ozone** layer. Thus their use is being gradually phased out.

Perhaps the most familiar form of the element is the compound stannous fluoride, a combination of fluorine and **tin**, which is used in toothpastes and dental treatments to prevent tooth decay. Fluoride, as it is better known, has also been added to the drinking water consumed by some 60 percent of Americans in communities across the country.

See also **Fluoride treatment, dental**

Folic acid

Folic acid is a member of the vitamin B family. This compound plays an important role in creating several components of the nucleic acids that are essential elements of all **cells**.

The human body's red and white blood cells appear to be particularly sensitive to a folic acid deficiency. So doctors often view **blood** disorders—such as megaloblastic **anemia** and sprue—as an early and important sign that vitamin B may be lacking. The major natural source of folic acid is green leafy vegetables. The acid takes its name from *folium,* the Latin word for leaf.

In 1945 Robert Angier and his coworkers at Lederle Laboratories identified the structure of folic acid and synthesized it (made it artificially).

Food chain

All organisms, dead and alive, are potential sources of energy and nutrition for other members of the environment.

There is no waste produced in a functioning, thriving natural ecosystem. All organisms, dead and alive, are potential sources of energy and nutrition for other members of the environment. For example, a worm digests tiny soil nutrients. Then a robin eats the worm. Next a wild cat eats the robin. When the cat dies, its body is broken down by bacterial decomposers such as worms. This process of exchanging energy and nutrients among living organisms is called a food chain.

The word "chain" is probably misleading, however, because it implies an orderly linkage of equal parts. Actual food chains are extraordinarily complex because there is no exact order specifying which creatures eat which others. It would be more accurate to consider each of the links as an energy carrier in a complex network of many interconnected food chains, called a food web.

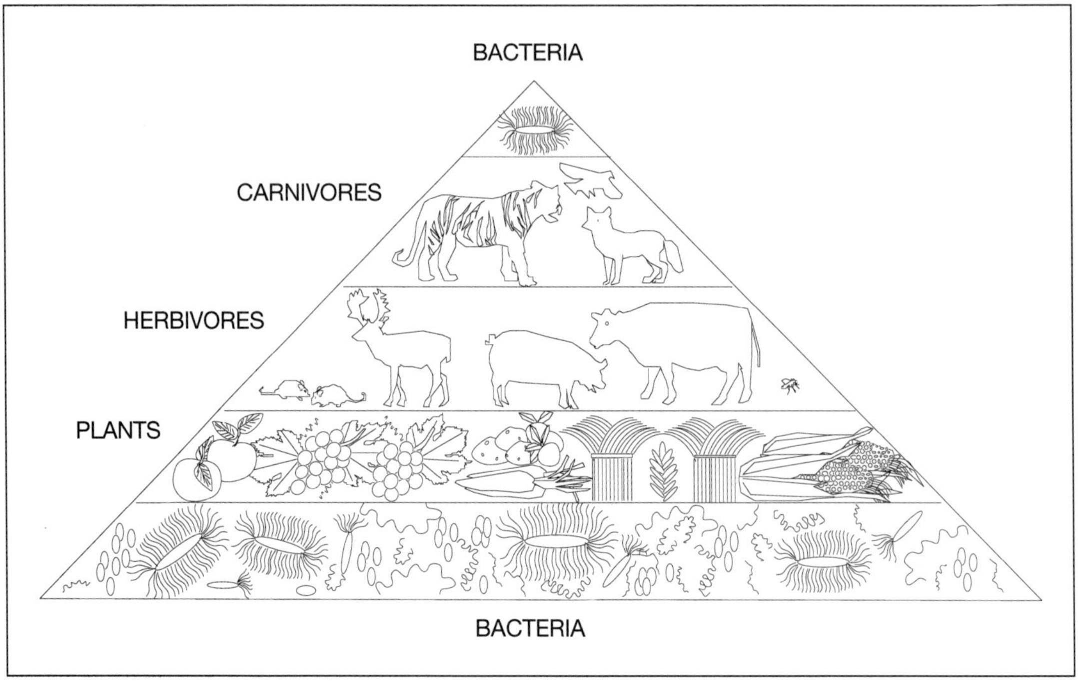

One way to picture a food chain is in the shape of a pyramid. The bottom layer consists of plants, which are eaten by herbivores, at the second level. In turn, these consumers become food for succeeding levels of flesh-eating animals.

Another way to picture a food chain is in the shape of a pyramid. The bottom layer of a food pyramid consists of the most abundant elements: plants that trap solar (sun) energy through **photosynthesis**. These plants are then eaten by herbivorous (plant-eating) animals, the primary consumers at the second level of the pyramid. In turn, these primary consumers become food for other animals, the secondary consumers at the third level of the pyramid. Each succeeding level consists of fewer, usually larger flesh-eating animals.

The consumers (such as humans) at the top of the pyramid do not represent the uppermost end of the food chain. When they die, they are eaten by tiny, microscopic organisms that serve as decomposers. When devoured by these **bacteria**, the nutrients are returned to the soil to be recycled into yet another food chain.

Food preservation

All foods begin to spoil as soon as they are harvested or slaughtered. Some spoiling is caused by such microorganisms as **bacteria** and mold. Other spoilage results from chemical changes within the food itself due to natural processes such as enzyme action or oxidation. The purpose of food preservation is to stop or slow down the spoilage.

Ancient Methods

Ages-old food preservation techniques include drying, salting, smoking, fermenting, pickling, cooling, and freezing.

Drying and Smoking

One of the most ancient methods of food preservation is sun- or air-drying. Drying works because it removes much of the food's water. Without adequate water, microorganisms cannot multiply and chemical activities greatly slow down. Dried meat was one of the earliest staple foods of hunters and people on the move. Once fire was discovered, prehistoric cave dwellers heat-dried meat and fish, which probably led to the development of smoking as another way to preserve these foods. The Phoenicians of the Middle East air-dried fish. Ancient Egyptians stockpiled dried grains. North American natives produced a nutritious food called pemmican by grinding together dried meat, dried fruit, and fat.

Bottles are sterilized prior to being filled with various juices at the Granicelli Cider Mill in Tivoli, New York. Sterilization kills the microorganisms that cause spoiling.

Modern methods of food preservation have helped decrease the incidence of starvation and disease.

Cooling and Freezing

Early Northern societies quickly learned that coolness as well as freezing helped preserve foods. Microbe growth and chemical changes slow down at low temperatures and completely stop when water is frozen. Pre-Columbian natives in Peru and Bolivia freeze-dried potatoes, while the early Japanese and Koreans freeze-dried their fish. Water evaporating through earthenware jars was used as a coolant in 2500 B.C. by Egyptians and East Indians. Ancient Chinese, Greeks, and Romans stored ice and mountain snow in cellars or icehouses to keep food cool.

Salting and Pickling

Salting, which also inhibits bacteria growth, was a preferred method of preserving fish as early as 3500 B.C. in the Mediterranean world, and was also practiced in ancient China.

Substances besides salt were found to slow food spoilage. The Chinese began using spices as preservatives around 2700 B.C. Ancient Egyptians used mustard seeds to keep fruit juice from spoiling. Jars of fruit preserved with honey have been found in the ruins of Pompeii, Italy. Melted fat—as Native Americans discovered with pemmican—preserved meat by sealing out air. Pickling—preserving foods in an acid substance like vinegar—was used during ancient times, too.

Fermenting

Fermentation was particularly useful for people in southern climates, where cooling and freezing were not practical. When a food ferments, it produces acids that prevent the growth of organisms that cause spoilage. Grapes, rice, and barley were fermented into wine and beer by early people. Fermentation also was used to produce cheese and yogurt from milk.

Canning

By the Middle Ages (A.D. 400-1450), all these ancient methods of preserving foods were widely practiced throughout Europe and Asia, often in combination. Salted fish became the staple food of poor people during this time—particularly salted herring, introduced in 1283 by Willem Beukelszoon of Holland. As the modern era approached, the Dutch navy in the mid-1700s developed a way of preserving beef in iron cans by packing it in hot fat and then sealing the cans. By the late 1700s, the Dutch also were preserving cooked, smoked salmon by packing it with hot butter or olive oil in sealed cans.

Modern Methods

Modern methods of food preservation include canning, mechanical refrigeration and freezing, the addition of chemicals, and irradiation.

With the Industrial Revolution (1760-1870), populations became concentrated in ever-growing cities and towns. Thus other methods were needed to preserve food reliably for transportation over long distances and for longer shelf life.

The crucial development was the invention of sophisticated canning techniques during the 1790s by the Frenchman Nicolas François Appert, who operated the world's first commercial cannery by 1804. Appert's method, which first used bottles, was greatly improved by the 1810 invention of the tin can in England. Used at first for Arctic expeditions and by the military, canned foods came into widespread use among the general population by the mid-1800s.

Packaged Frozen Foods

However, not all foods could be successfully canned. So reliable methods of refrigeration were needed. Icehouses were first used to store ice cut from frozen ponds and lakes. The 1851 invention of a commercial ice-making machine by American John Gorrie (1803-1855) led to the development of large-scale commercial refrigeration of foods for shipping and storage. Clarence Birdseye introduced tasty quick-frozen foods in 1925. Shortages of canned goods after World War II (1939-45) helped boost the popularity of frozen foods.

Food in containers ready to be frozen. Shortages of canned goods after World War II helped boost the popularity of frozen foods.

Dehydrated Foods

Modern methods of drying foods began in France in 1795 with a hot-air vegetable dehydrator. Dried eggs were widely sold in the United States after 1895, but dried food was not produced in volume here until it was used by soldiers during World War I (1914-18). World War II led to the development of dried skim milk, potato flakes, **instant coffee**, and soup mixes. After the war, freeze-drying was applied to items such as coffee and orange juice, and the technique continues to be applied to other foodstuffs today.

Chemical Preservatives

Chemicals are now commonly added to food to prevent spoilage. They include benzoic acid, sorbic acid, and sulfur dioxide. Antioxidants such as BHA and ascorbic acid (vitamin C) prevent compounds in food from combining with oxygen to produce inedible changes. Between 1925 and 1959, the African American chemist Lloyd A. Hall developed superior antioxidants and meat-curing salts that greatly improved many food products.

The use of chemical additives has not been without controversy. The spread of often unnecessary and sometimes harmful chemical additives to food during the late 1800s led to governmental regulation—the British Adulteration of Food and Drugs Act of 1875 and the U. S. Food and Drugs Act of 1906, among others.

Other Methods

Aseptic packaging is a relatively new way to keep food from spoiling. A food product is sterilized and then sealed in a sterilized container. Aseptic packages, including **plastic**, aluminum foil, and paper, are lighter and cheaper than the traditional metal and glass containers used for canning. Aseptically processed foods are sterilized much more quickly than foods in **cans** or bottles, so their flavor is better. Aseptic packaging became commercially available in 1981. However, controversy has developed about the amount of disposable containers produced by this method.

Irradiation is another relatively new method. Food is subjected to low doses of radiation to deactivate enzymes and to kill microorganisms and insects. Although public concern about the technique persists, certain irradiated foods, such as strawberries, were sold commercially in the United States in the early 1990s.

See also **Can and canned food**

Football

The vastly popular game of American football has its origins in the English sports of rugby and soccer. In the mid-1800s, a soccer-like game was popular in the eastern United States. The object of the game was to kick a round ball across the other team's goal line. Teams often consisted of 30 or more players.

As the game's popularity increased, stricter rules were created and schools organized teams. The National Collegiate Athletic Association (NCAA) contends that the first football game was played in 1869 between Rutgers and Princeton universities, but the rules of that game were more like those of a soccer match. The first official football game took place on May 14, 1874, between Harvard and McGill universities.

As more and more colleges began to play football seriously, blocking and tackling were introduced, and these skills soon became as important as running. Rules varied from game to game and teams simply agreed on them before playing.

As running, blocking, and tackling became more crucial to football, so did the players' strength and agility. Games became much more violent and, since no helmets or padding were worn, injuries soared. In 1905 President Theodore Roosevelt threatened to ban the game from the United States after seeing a particularly gruesome photograph of an injured player. Several colleges prohibited football because of such injuries.

In 1906, when forward passing became allowed, football games became less violent and more strategic, putting the emphasis on careful planning and organization.

In 1906, when forward passing became allowed, games became less violent and more strategic, putting the emphasis on careful planning and organization. Helmets, pads, and face guards better protected players from injury. Today, teams have 11 different players for offense and defense, and play on a standard field that is 100 yards long and marked at every fifth yard line.

Fractures, treatments and devices for treating

Beginning in the early 1980s, casts made of fiberglass plaster came into use.

In ancient times, broken bones could mean death for a victim. They often became infected and many times did not heal well, preventing the victim from working. Methods for healing fractures, or broken bones, were used to help the victim regain full use of an injured arm or leg.

In treating a fracture, the bone ends must first be brought back together. Then, the fracture must be held together until the bone ends naturally fuse. Closed or simple fractures, in which the bone ends do not penetrate the skin, were more easily treated. But open or compound fractures were usually fatal before antiseptics were introduced in the 1860s, because infection would set in.

Splints and Bandages

The earliest method for holding a reduced fracture in place were splints—rigid strips laid parallel to each other alongside the bone. Later, medical practitioners stiffened the **bandages** that held the splints in place, using wax or tree resin. In the sixteenth century the famous French surgeon Ambroise Paré (1510-1590) made casts of wax, cardboard, cloth, and parchment that hardened as they dried.

Casts and Traction

Splints remained the basic method of immobilization until 1852. Then Dutch army surgeon Antonius Mathijsen introduced roller bandages treated with quick-drying plaster of paris (gypsum). Broken bones could be held in place while the wet bandages were applied. When dry, the bandages became a rigid cast that held the bones perfectly in place during healing. As Mathijsen himself pointed out, Arab physicians had used plaster casts for centuries, but knowledge of the technique had not reached the West until the end of the eighteenth century.

Sterile Treatment of Wounds

In the mid-1800s, Joseph Lister introduced antiseptics to surgery, making possible the successful treatment of compound fractures. Compound fractures meant contamination of the wound, which almost always led to severe infection and usually resulted in death.

Since infection could not be avoided in pre-antiseptic days, the usual method of treating compound fractures until the late nineteenth century was **amputation.** (In the case of removal of the leg as high as the thigh, this often resulted in fatal infection as well.) Once Lister began the era of antiseptic surgery in 1865, infection in compound fractures could be controlled and these injuries could at last be treated successfully with surgery, casts, and plates.

Plaster of paris casts remained standard treatment for fractures for 130 years. Beginning in the early 1980s, casts made of **fiberglass** plaster came into use and today are valued for their light weight and water resistance.

In addition to splints and casts, fractures have been treated with extension and traction, which involve pulling on a broken limb with weights and pulleys to keep it straight during healing.

In 1893 Hugh Arbuthnot Lane of Great Britain introduced the use of steel screws to rejoin bones that would not heal together naturally. He improved the technique around 1905 by using steel plates screwed into the bone ends.

Frisbee

Did you know that the original Frisbee was actually a pie tin with the name "Frisbie" stamped on it? It's true—one of the world's most popular toys originated in a Connecticut bakery. After William Russell Frisbie's bakery opened in Bridgeport, Connecticut, in the 1870s, his pies and sugar cookies became popular. Years later in the 1920s, students at nearby Yale University started tossing his discarded pie tins and cookie tin lids and yelling "Frisbie." Just as college students of that era had popularized crazes such as eating goldfish, tossing pie tins became another fad.

The original Frisbee was a discarded pie plate.

About 20 years later in California, Walter Frederick Morrison used his basement laboratory to tinker with tenite, a **plastic** used in camera parts. The son of an inventor and the creator of the home popsicle maker, Morrison fashioned a plastic disk that could fly. Morrison improved the pie-tin shape by adding the familiar curled-under lip that keeps today's Frisbee more stable in the air.

Morrison named his flying disk "Li'l Abner" after the popular cartoon strip character. In the late 1940s, Morrison amazed the crowds at county fairs, carnivals, and beaches by tossing the Li'l Abner to his wife, claiming it was sliding along "invisible string" which he was willing to sell. The toy came free.

Morrison's invention became well-known enough to catch the attention of Spud Melin and Rich Knerr in 1957. This inventive pair made backyard slingshots, the sound of which gave them the idea for the name of their company—Wham-O. They bought rights to Morrison's toy and renamed it the "Pluto Platter" because of the nation's fascination with science fiction and flying saucers. The plastic spinning disk had the shapes of planets around the outside ring and the familiar grooves in the plastic. The Pluto Platter was a great success—especially on college campuses. A year later Wham-O changed the name to "Frisbee" to pay tribute to the original.

Today's Frisbees come in many colors and variations. International Frisbee championships are held, and there are even Frisbee-catching dogs. The military has tested Frisbees as a practical way of keeping flares up in the sky.

G

Gaia hypothesis

"Gaia" comes from the Greek word that means Goddess of Earth. The term has come to symbolize "Earth-Mother," or "Living Earth," meaning that Earth is a single living organism.

The Gaia concept is several hundred years old and evolved from the work of a few noted scientists. The hypothesis was first posed by the eighteenth-century Scottish geologist James Hutton, who referred to the Earth as a "Superorganism." Hutton viewed Earth's systems as codependent and existing within a single organism. Hutton later became known as the father of geology after publishing his *Theory of the Earth,* which pointed to volcanism (volcanic forces) as being the primary force shaping Earth.

In its modern form, the Gaia theory was put forward by James Lovelock, who published *Gaia: A New Look at Life on Earth.* Lovelock, an English chemist, suggested that Earth's biosphere acts as a single living system called Gaia. If left alone, the biosphere can regulate itself. Biosphere refers to Earth's land, sea, and air.

Lovelock drew on his background as a chemist to explain his theory. He explained that Earth provides a delicate balance between the atmospheric **carbon dioxide** and **oxygen** needed by living organisms. For Lovelock, Earth is responsible not only for creating this unique chemical makeup in the atmosphere, but also for other environmental characteristics that make life possible. For example, he argued that it is no accident that the level of oxygen is kept remarkably constant in the atmosphere at 21 percent.

James Lovelock with a device he invented to measure chlorofluorocarbons (CFCs) in the air. Lovelock is a proponent of the Gaia theory, which suggests that Earth's biosphere acts as a single living system that can regulate itself.

Lovelock explained that the biosphere can create the environment that promotes its own stability. He warned that by tampering with Earth's own environmental balancing mechanisms, we are placing ourselves and our planet at grave risk. He points to global warming and **ozone** depletion as indications of this risk.

Both sides of the earth conservation argument use Lovelock's arguments. Radical environmentalists insist that human practices are upsetting Earth's ability to regulate itself. Industry representatives, on the other hand, argue that Earth can continue to survive based on its history of self-preservation through adaptation.

Galileo Galilei

Italian mathematician and astronomer Galileo Galilei (1564-1642) changed forever our notion of how we fit into the universe. He is the father of the modern experimental method. Before Galileo, scientists and thinkers had no systematic way of looking at the physical world. Their method for coming to conclusions was a combination of hypothesis and guesswork, with some basis in logic and observation but often wholly based on religious authority or simply irrational prejudice.

In contrast, Galileo introduced the practice of proving or disproving a scientific theory by conducting tests and observing the results. By this procedure he made numerous significant discoveries, particularly in the fields of physics and astronomy. Physics is the science that deals with matter, energy, force, and motion. Astronomy is the study of the universe. Galileo's approach came to be called the experimental method.

Student

The son of a musician, Galileo was born in Pisa, Italy. He received his early education at a monastery near Florence, and in 1581 entered the University of Pisa to study medicine. While a student he observed a hang-

Medici Princes

In Galileo's time, many famous artists and thinkers were supported by the aristocracy. In recognition of their sponsorship of his studies, Galileo named the moons of Jupiter *Sidera Medicea* ("Medicean stars"). Thus he honored Cosimo de Medici, the Grand Duke of Tuscany, whom Galileo served as "first philosopher and mathematician" after leaving the University of Pisa in 1610.

ing lamp that was swinging back and forth. Galileo noted that the amount of time it took the lamp to complete a swing remained constant, even when the distance of the swing became smaller. He later experimented with other suspended (hanging) objects and discovered that they behaved in the same way. What Galileo had witnessed was the principle of the pendulum, which he later applied to regulating clocks.

While at the University of Pisa, Galileo listened in on a geometry lesson and afterward abandoned his medical studies to devote himself to mathematics. However, he was unable to complete a degree at the university due to lack of funds.

He returned to Florence in 1585, and the following year published an essay describing his invention of the hydrostatic balance. This device determined the specific **gravity** of objects by weighing them in water. With this invention Galileo gained a scientific reputation throughout Italy.

Teacher

In 1589 Galileo obtained a professorship in mathematics at the University of Pisa, where he remained for the next 18 years. During this time Galileo conducted his most important scientific research and published papers documenting his work. Among these was an untitled treatise in which he discredited Greek philosopher Aristotle's contention that heavier objects fell at a faster rate than lighter ones, an idea that until that time had been universally accepted. Galileo offered theoretical proof that falling bodies accelerate (speed up) or decelerate (slow down) uniformly with time, a principle that was later called the law of uniformly accelerated motion.

Galileo Galilei

Astronomer

In 1609 Galileo first learned of the telescope, which had recently been invented by a Dutch lens-grinder, Hans Lippershey. After improving on the original design, Galileo constructed a telescope with a magnifying power of 32, enabling him to make a number of important astronomical discoveries.

With his telescope Galileo found that the **Moon** was not a perfectly smooth sphere, as was previously thought, but irregular in its surface. He also observed that the Milky Way was composed of individual stars. Galileo's study of **Jupiter** resulted in his discovery of four of its moons, which he later called "satellites," a term suggested by the German astronomer Johannes Kepler.

Galileo Galilei used his telescope to probe the universe and forever changed man's notion of his importance in the grand scheme of things.

Training his telescope on **Venus**, Galileo discovered that this planet exhibited phases much like the moon, and for the same reason: Venus did not produce its own light but was illuminated by the **Sun.** In addition Galileo noted that **Saturn** was encompassed by rings, which his telescope allowed him to see only as "protuberances" (bumps) on either side of the planet. Galileo also observed sunspots.

Author

In 1613 Galileo published the book that was to prove his undoing. For the first time Galileo was publishing evidence for and openly defending the model of the **solar system** earlier proposed by the Polish astronomer Nicolaus Copernicus. In 1531 Copernicus argued that Earth, rather than being positioned at the center of the universe, was only one of several galactic bodies that orbited the Sun.

While there was some support for Galileo's proof of the Copernican theory, Roman Catholic officials disagreed with his view of the universe. The Church was very powerful at the time. In 1616 it published a decree that declared the Copernican system "false and erroneous," and Galileo was told to reject the theory and his book. For several years thereafter Galileo led a quiet existence at his home near Florence.

Galileo wanted to have the decree canceled, and in 1624 he traveled to Rome to make his appeal to the newly elected pope, Urban VIII. The pope would not revoke the decree but did give Galileo permission to write about the Copernican system. In addition, the pope demanded that Copernicus's theory would not be given preference to the church-sanctioned model that held that Earth was the center of the universe.

Heretic

With Urban's imprimatur (seal of approval), Galileo wrote his *Dialogue Concerning the Two Chief World Systems—Ptolemaic and Copernican,* which was published in 1632. Despite his agreement not to favor the Copernican view, his writing in the *Dialogue* made the objections to the Sun-centered model unconvincing and even ridiculous.

Galileo was summoned to Rome to stand before the Inquisition—a group of priests whose job it was to seek out and punish heretics (those who did not support the Church's beliefs). Galileo was accused of violating the papal order of 1616 that forbade him to promote the Copernican theory. Galileo was put on trial for heresy, found guilty, and ordered to recant his errors. At some point during this ordeal Galileo is supposed to have made his famous statement: "And yet it moves," referring to the Copernican doctrine of Earth's rotation on its axis.

While the judgment against Galileo included a term of imprisonment, the pope commuted this sentence to house arrest at Galileo's home near Florence. There Galileo continued to work, although he was forbidden to publish any further works. He died in 1642.

Game theory

The theory of games had its modern origins in a paper presented by the Hungarian-American mathematician John von Neumann before the Mathematical Society of Göttingen, Germany, in 1928. In his paper, von Neumann outlined a mathematical system for determining the best strategy for a player who wants to get good results with few losses in some type of contest. Von Neumann analyzed the ways in which players could make choices so that they could achieve the best possible "pay-off" after a given period of time in the game.

Early Thinkers

The application of mathematics to games and other types of contests can be traced at least to the seventeenth century. French mathematician Abraham de Moivre was one of the first scholars to apply the mathematics of probability to practical situations. He developed formulas for calculating the probability of various events, and was hired as a consultant by gambling syndicates and insurance companies.

The application of mathematics to games and other types of contests can be traced at least to the seventeenth century.

Some authorities give credit not to von Neumann, but to the French mathematician Émile Félix-Édouard-Justin Borel as founder of modern game theory. Between 1921 and 1927, Borel published a series of papers analyzing many aspects of game theory. He defined games of strategy and developed theorems (predictions) dealing with symmetric games, infinite games, optimal strategies, and three-, five-, and seven-player games. He also examined the application of game theory to economics and warfare.

Basic Concepts

Von Neumann deserves credit for formulating many fundamental concepts in game theory. He spent a major part of his life, from 1928 to his death in 1957, on the development of this field of mathematics. In 1944 he wrote his culminating work on the topic, *Theory of Games and Economic Behavior,* with economist Oskar Morgenstern.

Game theory has become enormously important in many practical situations today. Military strategists use its mathematical theorems, for example, to estimate various possible responses that two or more opposing forces might use in various situations.

Gamma ray

By 1900 it became apparent that nuclear radiation consisted of at least two parts, alpha rays and beta rays. Gamma rays are released during alpha and/or beta decay. The rays are apparently produced as a result of changes in nuclear energy levels, just as **X-rays** are liberated during changes in electron energy levels.

Gamma rays were discovered in 1900 by the French physicist Paul Ulrich Villard. Villard discovered that this third form of radiation was unaffected by a **magnetic field** and was even more energetic than beta rays. Whereas alpha rays are stopped by a few centimeters of air and beta rays by a few centimeters of aluminum, Villard's new rays could only be stopped by a relatively thick piece of lead.

The nature of gamma rays was a topic of considerable dispute. Some scientists thought that they were composed of very energetic particles, as was the case with alpha and beta rays. Others suggested that they were a form of **electromagnetic wave**, similar to X-rays. A 1914 experiment showed that gamma rays were a form of electromagnetic radiation even more energetic than X-rays.

In fact, there is no difference between high energy X-rays and low energy gamma rays. As a matter of convenience, the term "X-rays" is reserved for radiation that originates outside the atomic nucleus and "gamma rays" for radiation that originates within it.

Gamma ray astronomy

Satellites observe gamma ray energy to learn how stars are formed.

Astronomers have been observing the visible portion of the electromagnetic spectrum for centuries, but this is just a very small "window" to the universe, and it provides a limited view. With advanced technology, astronomers can now observe the nonvisible parts of the spectrum, including **X-rays** and radio waves.

One phenomenon under observation is **gamma rays**. Better understanding of gamma rays will help astronomers establish a better picture of energy production in stars and stellar evolution. The **Sun** emits as much energy in a single second as humans have consumed throughout history, and the understanding of stellar energy could lead to more efficient methods of energy production on Earth.

Gamma rays are high-energy particles. They were first detected on Earth through the decay of radioactive elements but are also produced by nuclear reactions. The most efficient nuclear reactors are the stars, and they produce gamma rays deep in their interiors. By observing this radiation, astronomers can "see" into stellar cores.

Satellite Data

Gamma rays from space do not penetrate to the surface of the earth because the **ozone** layer absorbs high-energy radiation. If it did not, life on this planet would not last long. However, this protective atmosphere is a barrier to the study of cosmic rays. To detect them, orbiting satellites must be used.

Cosmic gamma rays were first discovered in 1967 by U.S. satellites. These Vela satellites had been launched to monitor nuclear bomb explosions on Earth, but they detected gamma ray bursts that came from outside the **solar system**. Other satellites found bursts lasting from as short as one-tenth of a second to several seconds. The sources seem to be scattered randomly across the sky.

The first satellite devoted to studying cosmic rays, U.S. *Explorer 48* (*SAS B*), was launched November 16, 1972. In the late 1970s, NASA launched three *High-Energy Astronomical Observatories* (*HEAOs*). *HEAO 1* and *HEAO 2* (the Einstein Observatory) operated in the X-ray wavelength. *HEAO 3* made cosmic ray and gamma ray observations. European satellites have advanced knowledge as well.

Much higher-energy gamma radiation was detected by *OSO 3*. It found radiation along the plane of the Milky Way galaxy. Later, *SAS 2* (*Small Astronomy Satellite*) found the Milky Way looks much the same in gamma rays as it does in visible light. Other gamma ray sources, such as the Crab Nebula, have been identified as old supernovas.

Gas mask

Anselme Payen and Garrett Augustus Morgan are credited with inventing the modern gas mask. This device filters air to protect the wearer from breathing poisoned fumes or smoke.

In 1822 Payen, a sugar manufacturer, used animal charcoal to remove large-molecular impurities from his product. Eventually his use of charcoal to absorb impurities became an important feature of the gas masks used in World War I (1914-18).

Morgan, an African American, was born in Paris, Tennessee, in 1877. In 1912 he came out with his most important invention, the Safety Hood or "Breathing Device," as he called it. His device was a hood placed over the head of the user. Two tubes were connected to the hood: one provided fresh air while the other took away used air. The fresh-air tube was lined with an absorbent material that could be moistened with water to keep out smoke and dust particles.

While Morgan demonstrated his gas mask at several exhibits and fairs, it was not until later that his invention gained widespread interest. In 1916 a violent explosion at the Cleveland Waterworks trapped workers inside a tunnel under Lake Erie. Heavy smoke and poisonous gases prevented any rescue attempts until Morgan arrived with several of his gas masks. He and three volunteers used the masks to save the lives of 32 men by carrying them out of the tunnel.

Meanwhile, during World War I the German army used poison gas for the first time in battle. At first the English used chemically treated cotton pads tied over their mouths and noses. Soon they were building advanced gas masks based on Morgan's invention. These large devices consisted of a mask and a large tube that connected the mask to a canister that hung in front of the soldier's body. Inside the canister was charcoal, which filtered the poison gases.

By World War II (1939-45), gas masks were lighter, better fitting, and designed to allow better vision. The filter was redesigned to wear over the shoulder for easier carrying.

In the 1960s, the United States military developed the M-17 mask that proved to be a breakthrough in protecting soldiers from biological,

American troops wearing gas masks advance in France during World War I, May 20, 1918. The soldier at left, unable to don his mask, clutches his throat as he breathes in poisonous gas.

Gas mask

A firefighter wearing a gas mask, October 19, 1963. Today gas masks are primarily used as protection for firefighters and law enforcement officers and in such environments as chemical plants and mines.

chemical, and radiation agents. The M-17 had no hose or external canister. In the M-17, air was filtered through pads of flexible material enclosed in cavities molded into the facepiece of the mask. Today gas masks are primarily used in such environments as chemical plants and mines, and as protection for firefighters and law enforcement officers.

Gasoline

Premium or regular? Leaded or unleaded? These are the main choices at the gas station pump today. But gasoline is actually a much more complex substance than these labels suggest. Gasolines are carefully mixed at the oil refinery to produce specific blends, so that gasoline sold at a service station in Minnesota is very different from that sold in Florida.

Gasolines are carefully mixed at the oil refinery to produce specific blends, so that gasoline sold at a service station in Minnesota is very different from that sold in Florida.

Oil refiners blend gasolines to make car engines run better in different climates and different seasons. During the winter, gasoline must vaporize (turn to gas) faster so cars are easier to start on a cold day. During the summer, oil refiners produce gasoline blends that are harder to vaporize, because bubbles of vapor in the fueling system can cause vapor lock during hot weather. The bubbles prevent gas from moving through the fuel line to the engine.

Early Fuels

The gasoline used by **automobiles** in the early 1900s, however, was a much simpler product. Before then, crude oil was distilled and separated to produce kerosene fuel for oil lamps. Gasoline that was produced along with the kerosene was discarded because no one had discovered a use for it.

When the automobile was invented, that industry created a new market for gasoline. At first, automobile engines used "straight-run" gasoline—the natural gasoline produced by distilling crude oil. But this process yielded fewer than 15 barrels of gasoline from each barrel of oil. After the mass production of cars began in 1908, oil refiners could not keep up with the growing demand for gasoline.

In 1913, just in time for World War I (1914-18), a process was invented to increase the amount of gasoline produced from crude oil. William Burton, who worked for Standard Oil of Indiana, developed thermal cracking, a process that uses heat and pressure to better reduce oil to gasoline.

During World War II (1939-45), other new refining processes greatly increased the United States' output of gasoline. More than 80 percent of the aviation fuel used by the Allies (United States, England, and France) during the war was supplied by the United States. In Europe, gasoline became extremely scarce, and the German army had to rely on cruder types of gasoline that were produced from coal and heavy oil.

Gasoline can be made from just about any substance containing hydrogen and carbon.

Gasoline Ingredients

Gasoline can be made from just about any substance containing **hydrogen** and **carbon**. Today's gasolines are blended from hundreds of hydrocarbons, and different combinations are produced to meet the needs of different engines and to prevent engine knock.

In the early 1900s, engine knock was recognized as a problem, and the auto industry began searching for a fuel that could work without knocking. While engineers experimented with different engine designs, chemists explored "additives"—substances that could be added to gasoline to prevent knocking. In 1921 a team of American chemists led by Thomas Midgley, Jr., and T. A. Boyd made a spectacular breakthrough at General Motors.

The essential compound turned out to be tetraethyl lead. When added to gasoline in tiny amounts, tetraethyl lead prevents engine knock and increases the gasoline's octane rating. Unfortunately, tetraethyl lead pollutes the air with poisonous lead compounds when the gasoline is burned, and today leaded gasoline is being phased out.

Instead, new engines have been designed to run on lower octane gasoline, which is made of **hydrocarbons** that are resistant to knock. New additives have also been formulated to improve unleaded gasoline. Other modern additives preserve fuel quality and prevent rust, ice, and deposits of burned solids in the engine and fueling system.

During the 1970s, leaded gasoline became associated with another problem. The lead in the gasoline exhaust fumes was ruining the car's antipollution equipment. In 1970 the government required automakers to sharply reduce the amount and number of toxic (poisonous) emissions. To meet these standards, automakers introduced catalytic converters—devices that are attached to the exhaust system just behind the exhaust manifold. Most converters turn harmful **carbon monoxide** and hydrocarbons into harmless **carbon dioxide** and water vapor. The catalysts are easily poisoned by lead, however, which clogs their reactive surfaces. That is why most cars must now use unleaded gasoline.

Oil Shortages

Plentiful gasoline has made many aspects of life convenient and pleasurable. People can travel longer distances to get to their jobs or to go on vacation. Farmers can produce more food by using gasoline-fueled machinery.

But by the early 1970s, gasoline use had grown so enormously that oil refiners began to turn to imported oil. When foreign oil supplies were disrupted, gasoline supplies suddenly became limited. People waited for hours to fill up their tanks at the service station, and gasoline prices skyrocketed from less than 40 cents to more than a dollar a gallon.

Since then automakers have introduced smaller cars that weigh less and thus use less fuel. It was the gasoline shortages of the 1970s that made compact Japanese cars popular in the United States for the first time. For a time the government also encouraged people to conserve gasoline by using public transportation, and states reduced highway speed limits.

During the 1980s, these conservation measures were neglected, and higher speeds are now allowed on some stretches of highway. However, the government still requires automakers to continue increasing fuel economy and reducing pollutant emissions.

Besides giving off exhaust fumes, gasoline can evaporate into the air while it is being pumped into a car's tank. Service stations will soon be required to control these vapors with anti-evaporation equipment.

Gasoline can also leak from underground storage tanks, which lie below nearly every service station and distribution facility. Gasoline is an explosive, toxic contaminant and has become a major contributor to groundwater pollution.

See also **Automobile, gasoline; Internal combustion engine; Oil refining**

Gene

Genes are the physical units of **heredity**. They are located along each **chromosome** in the **cells** of the human body. For each physical trait—eye color, height, hair color—a person inherits two genes or two groups of genes, one from each parent. Because both of these genes cannot be expressed together, one usually overpowers the other. The more powerful gene is called the dominant gene and the weaker is called the recessive gene.

Genes are the units of heredity that give us our hair color, height, and intelligence. They make us who we are.

All genes on the same chromosome are called linked genes because they are usually inherited together. (For instance, the genes for red hair, freckles, and fair skin may all appear on the same chromosome.) Genes on the X and Y chromosomes are called sex-linked because the X and Y chromosomes are the ones that determine sex. (Men have an XY pair of chromosomes while females have an XX pair.)

Sometimes genes on the same chromosome are not inherited together. When sex cells divide to form an egg or sperm cell (a process called meiosis), each chromosome pairs off with a partner. As the chromosomes lie side by side, groups of genes from one chromosome may trade places with groups of genes from the partner chromosome. This is

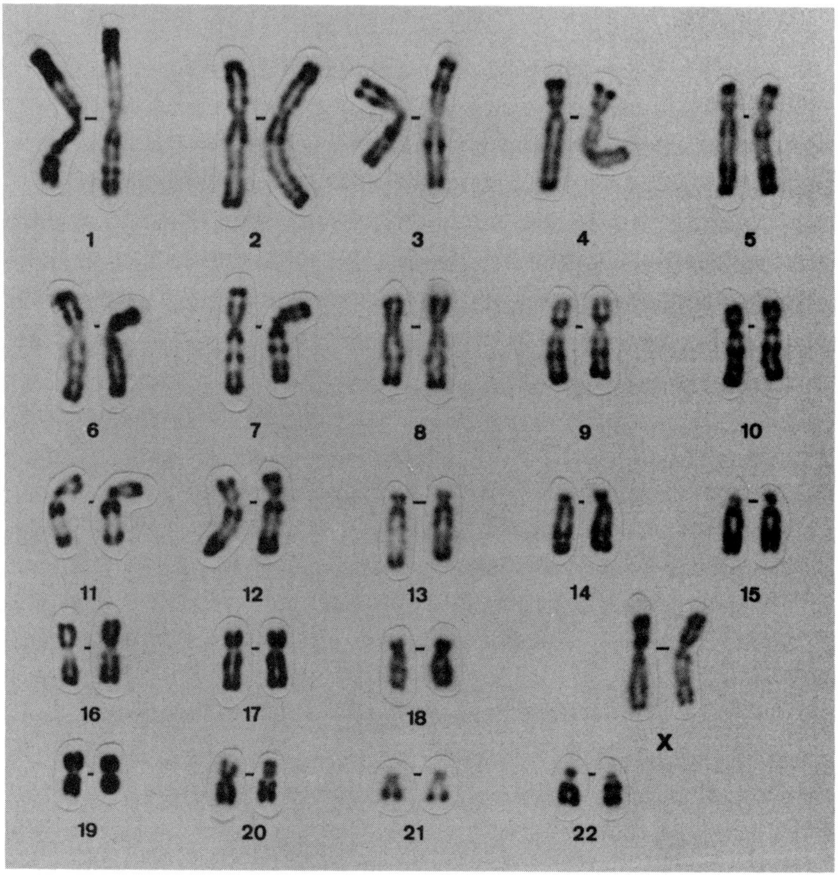

Healthy female chromosomes. Genes are located along each chromosome in the cells of the human body. For each physical trait a person inherits two genes or two groups of genes, one from each parent.

called crossing over and thus explains how families inherit different combinations of linked traits.

In 1910 Thomas Hunt Morgan began to uncover the interesting relationship between genes and chromosomes. Morgan was the geneticist who discovered that genes are located on chromosomes and that genes are linked. But as Morgan and his colleagues worked with more and more characteristics simultaneously, they discovered the crossover phenomenon. Morgan and his colleagues went on to develop and perfect these and other gene concepts, which laid the groundwork for all genetically based medical research.

Mendel's Contributions

Austrian monk and botanist Gregor Mendel (1822-1884) introduced the world to hereditary factors—genes—that determine all hereditary traits. During his experiments with pea plants, Mendel noticed that the plants inherited traits in a predictable way. It was as though the pea plants had a pair of factors responsible for each trait. Even though he never actually saw them, Mendel was convinced that tiny independent units determined how an individual would develop. Before Mendel's findings, traits were thought to be passed on through a mixing of the mother and father's characteristics, much like a blending of two liquids.

When Mendel's laws of heredity were rediscovered in 1900 they became vitally important to biologists. Among other things, Mendel's laws established heredity as a combining of independent units, not a blending of two liquids. Wilhelm Johannsen, a strong supporter of Mendel's theories, coined the term "gene" to replace the variety of terms used to describe hereditary factors. His definition of the gene led him to distinguish between genotype (an organism's genetic makeup) and phenotype (an organism's appearance).

See also **Evolutionary theory**

Thomas Hunt Morgan was the geneticist who discovered that genes are located on chromosomes and that genes are linked.

Gene therapy

Gene therapy is the treatment of disease with genes that have been engineered for that specific purpose. (See **Genetic engineering** for a description of how this is done.)

An altered gene that has been returned to a patient's body can save the person's life.

The first human gene therapy was approved for clinical trial in the United States in May 1989. At the end of 1992, at least 37 gene therapy projects were completed, in progress, or approved in China, France, Italy, the Netherlands, and the United States. Each country has its own approval process, designed to protect the patient, the health workers, and the public.

In the United States, each procedure must be approved by the National Institutes of Health Recombinant DNA Advisory Committee, by the U.S. Food and Drug Administration, and by the director of the National Institutes of Health.

The two most extensive gene therapy trials so far have been on severe combined immunodeficiency (SCID) and malignant melanoma. SCID is a rare disease that keeps the person's immune system from functioning. It was well-publicized in the case of a teenager named David who lived for several years in a plastic bubble to protect him from infection.

Some instances of SCID result from a genetic mutation that prevents production of the protein adenosine deaminase (ADA), which protects immune system white cells called lymphocytes. In September 1990, R. Michael Blaese and W. French Anderson at the U.S. National Institutes of Health performed the world's first gene therapy on a four-year-old with SCID. A normal gene for ADA was inserted into a virus and allowed to enter lymphocytes that were withdrawn from the child's body. Then the girl was injected with the altered cells.

During the next eighteen months, the girl had several series of injections, along with other treatment. A second patient, a nine-year-old girl, had similar treatments. As expected, the cells encouraged production of ADA in both children, allowing them to go to school and have only the normal number of infections. There were no side effects.

Similar treatments have been used on children in other countries.

Melanoma is a type of skin cancer that is often fatal. Again, doctors withdraw a sample of the patient's own cells, insert an altered gene, and return the new cells to the patient. The purpose of this procedure with melanoma patients is to introduce a protein that will kill the tumor.

A genetic treatment for cystic fibrosis, a lung disease, was approved in 1992. The therapy calls for inserting a needed gene into an inactive cold virus that the patients inhale. If the gene enters the lung and functions, it may prevent the production of the mucus that blocks a patient's breathing.

In familial hypercholesterolemia, patients lack a gene for disposing of harmful low-density lipoprotein **cholesterol**, allowing it to build up in their bodies. People lacking both copies of the gene usually die from a heart attack in their early teens. Someone with only one copy suffers from severe coronary (heart) disease. Scientists at several medical centers are studying insertion of the needed gene into cells from a patient's liver, then injecting the cells into the person's body.

The bleeding disease hemophilia B occurs in people whose blood lacks clotting factor IX (nine). Scientists in China are attempting to engineer cells with this factor.

Studies are also underway on genetic therapy for **AIDS** (**Acquired Immune Deficiency Syndrome**), liver failure, leukemia, brain tumors, and lung cancer.

Genetically engineered blood-clotting factor

Excessive, uncontrolled bleeding can be fatal. One well-known disease associated with uncontrolled bleeding is **hemophilia** (he-mo-fil-e-a). Most hemophiliacs bleed uncontrollably because a single gene on the X chromosome lacks the instructions that tell the cell how to make a specific protein. This protein, called factor VIII (8), is required for blood to clot normally.

Purified factor VIII extracted from human blood became available around 1960, but it was very expensive. Furthermore, viral impurities in the human factor VIII placed many hemophiliac patients at risk of contracting serious diseases, including **hepatitis** and, later, **AIDS** (**Acquired Immune Deficiency Syndrome**).

Artificially Creating Factor VIII

In the early 1980s, scientists at Genentech and Chiron Corporation in California and at the Massachusetts-based Genetics Institute began developing **genetic engineering** techniques to obtain pure, inexpensive factor VIII artificially. Genetic engineering refers to human methods for rearranging genes—removing or adding them or transferring them from one organism to another.

At Genentech, Richard Lawn, Gordon Vehar, and their coworkers succeeded in isolating the normal gene for factor VIII in healthy people. They inserted the gene into laboratory-grown hamster cells, where it joined with the **DNA** (**deoxyribonucleic acid**) of the hamsters. The hamster cells then used the genetic instructions in the DNA to make pure human factor VIII.

In April 1984, after many months of work, tests showed that the genetically engineered factor VIII is able to clot hemophiliac blood. This promising method of treating hemophilia was inexpensive and safe. However, it had one problem: it was difficult to control the amount of factor VIII that the cells produced, and too much factor VIII caused the blood to stop circulating properly.

Although it will take several more years of work before the gene itself can be introduced directly into a patient, tests are currently underway to determine the best dosage of artificial factor VIII for hemophiliac patients.

Genetic code

Studying the genetic code reveals which genes are responsible for inherited diseases.

By the early 1950s, scientists knew that **genes** were made of **deoxyribonucleic acid (DNA)** and that specific proteins were the products of specific genes. The exact link between DNA and proteins was less well understood, however. Since proteins are considered the language of life, researchers believed that the DNA molecule, with its four nitrogenous bases, might be the code for this language. This is how the term "genetic code" originated.

Protein molecules are comprised of **amino acids**. There are 20 biologically important amino acids. Only four different bases, including adenine (A), thymine (T), cytosine (C), and guanine (G), are found in DNA. When each of these bases combines with a sugar and a phosphate molecule, a nucleotide unit is formed. How could only four different nucleotides code for 20 different amino acids? Scientists reasoned that if a single nucleotide coded one amino acid, only four amino acids could be provided for. If two nucleotides specified one amino acid, then there could be a maximum number of 16 possible arrangements.

George Gamow, a Nobel Prize winner, demonstrated that at least three nucleotides in sequence were required to code for a single amino acid. This would provide for 64 possible combinations or codons—more than enough different "messages" to code for the 20 amino acids.

By the 1960s, the mystery of the genetic code had been solved. The 20 amino acids are coded by 61 triplet codons. Three additional codons do not code for any amino acids but direct the cell as to when it should cease protein synthesis.

Today we can make synthetic or artificial genes, and we can change genes and observe the results of those changes. This ability to change or manipulate genes has been a valuable tool in studying genetic disorders and the mechanisms of cancerous cells. Artificial genes are now used to obtain large amounts of valuable proteins for human dietary and medical needs.

Genetic engineering

Man's manipulation of genes, the building blocks of life, has raised serious moral questions.

Genetic engineering is the human altering of the genetic material of living cells. The purpose is to make them capable of producing new substances or performing new functions. The technique became possible during the 1950s when scientists discovered the structure of **DNA** (**deoxyribonucleic acid**) molecules and learned how those molecules store and transmit genetic information.

How does the process work? Suppose that the base sequence T-G-G-C-T-A-C-T on a DNA molecule carries the instruction "make insulin." (The actual sequence for such a message would in reality be much longer.) DNA in certain cells in the pancreas would normally contain that base sequence since that is where insulin is produced in mammals.

But that base sequence carries the same message no matter where it is found. If a way could be found to insert that base sequence into the DNA of bacteria, for example, then those bacteria would be capable of manufacturing insulin.

Human Applications

The possible applications of genetic engineering are nearly limitless. For example, it is now possible to produce a number of natural products that were previously available in only very limited amounts. Until the 1980s, for example, the only supply of insulin available to diabetics was animals slaughtered for meat or other purposes. That supply was never adequate to treat all diabetics at moderate cost. In 1982, however, the U.S. Food and Drug Administration approved insulin produced by genetically altered organisms, the first such product to become available.

The potential commercial value of genetically engineered products was not lost on entrepreneurs in the 1970s. A few foresighted individuals believed that recombinant DNA would transform American technology as computers had in the 1950s. In many cases, the founders of the first genetic engineering firms were scientists themselves, often those involved in basic research in the field.

Genetic engineering (or, more generally, biotechnology) firms have, however, long been a source of controversy. Many question whether individual scientists have the right to make a personal profit by opening their own companies that are based on research carried out at public universities and paid for with federal funds. As of the early 1990s, working relationships had, in many cases, been formalized among universities, individual researchers, and the corporations they establish. But not everyone is satisfied that the ethical issues involved in such arrangements are settled. Ethics is the system of moral principles that a person uses to govern his or her life and decide what is right and wrong.

One of the most exciting potential applications of genetic engineering involves the treatment of genetic disorders. Medical scientists now know of about 3,000 disorders that arise because of errors in an individual's DNA. Conditions such as sickle-cell anemia, Tay-Sachs disease, Duchenne **muscular dystrophy**, Huntington's chorea, **cystic fibrosis**, and Lesch-Nyhan syndrome are the result of the loss, mistaken insertion, or change of a single nitrogen base in a DNA molecule.

The techniques of genetic engineering make it possible for scientists to provide individuals who lack a certain gene with correct copies of that gene. If and when that correct gene begins to function, the genetic disorder may be cured. This procedure is known as human gene therapy (HGT).

But many critics worry about where HGT might lead. If we can cure genetic disorders, can we also design individuals who are taller, more intelligent, or better looking? Will humans know when to say "enough" to the changes that can be made with HGT? These are some of the ethical questions that surround genetic engineering.

Agricultural Applications

Genetic engineering also promises a revolution in agriculture. It is now possible to produce plants that will survive freezing temperatures, that will take longer to ripen, that will convert atmospheric nitrogen to a form they can use, that will manufacture their own resistance to pests, and so on.

By 1988 scientists had tested more than two dozen kinds of plants engineered to have special properties such as these.

Many other applications of genetic engineering have already been developed or are likely to be realized in the future. In every case, however, the glowing promises of each new technique are balanced by the new social, economic, and ethical questions that are raised.

Genetic fingerprinting

Fingerprints are unique to each individual. Methods of recording and matching fingerprints have allowed police to correctly identify many criminals. Genetic scientists have recently developed another tool for identification based on the uniqueness of each person's **genes**.

A genetic fingerprint. Because the genetic fingerprint of each person is unique, genetic fingerprinting has helped to link suspects to crimes where a single drop of blood was the only clue.

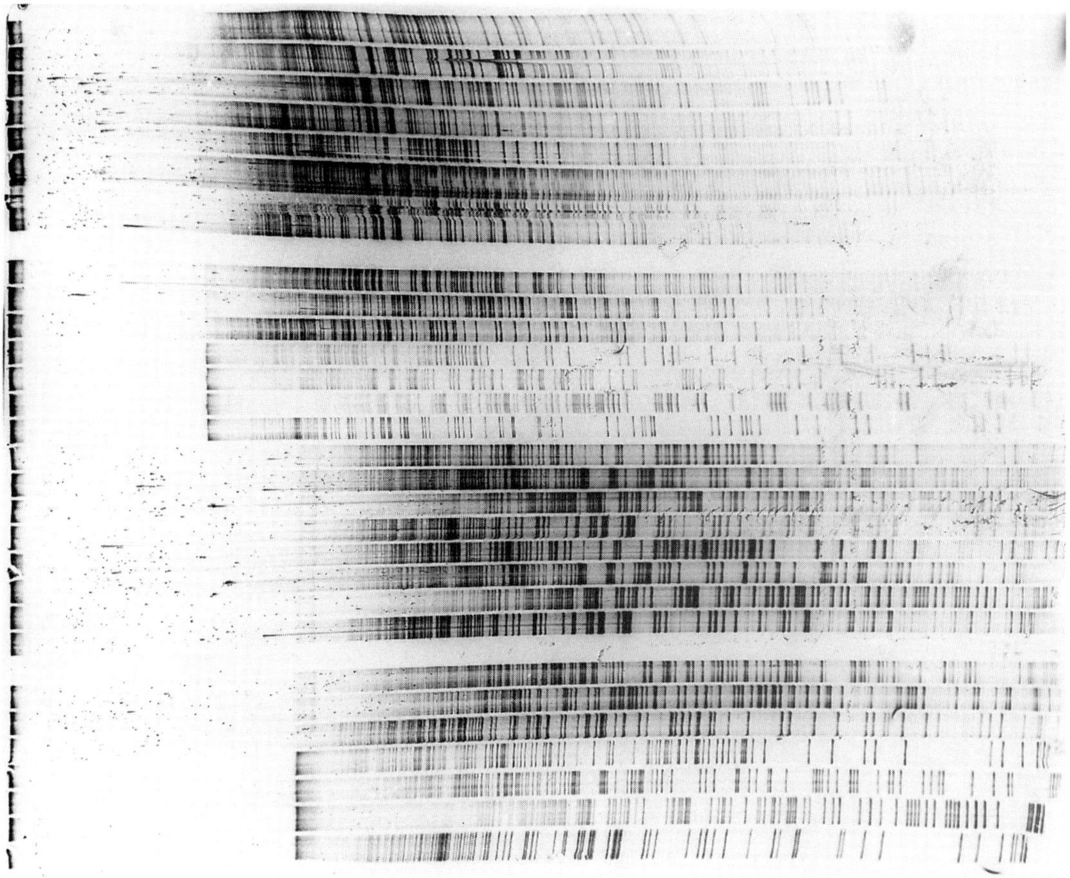

Genetic differences between people account for the large variations we see between individuals. Each human has approximately 100,000 genes in the chemical form of **DNA** (**deoxyribonucleic acid**). The genetic information coded in the genes varies greatly between individuals. Thus, no two humans, except for identical twins, have exactly the same **genetic code**. A description of a person's DNA that is detailed enough to distinguish it from another person's DNA is called a DNA or genetic "fingerprint."

In 1985 an English researcher named Alec Jeffreys developed a technique to visualize a person's genetic code. This direct DNA analysis revealed so much variation in the genetic code between different people that even a small section of the entire genetic code could identify an individual's special combination of traits.

Three years later, Henry Erlich developed a method of DNA fingerprinting so sensitive that it could be used to identify an individual from an extremely small sample of hair, blood, semen, or skin. Erlich's technique used Jeffreys's traditional method and combined it with a new technique. Using his new method, Erlich was able to duplicate and heat-separate the DNA fragments from a single human hair root many times. The amplified DNA was then used to obtain a DNA fingerprint.

Genetic fingerprinting has already proved to be a very useful tool. Initially, it was used exclusively in forensic (criminal) science and law. This technique has helped to link suspects to crimes where a single drop of blood was the only clue. Maternity and paternity matters have also been settled using genetic fingerprinting.

Geodesic dome

The geodesic dome was developed by American architect R. Buckminster Fuller around 1949. A dome is a round building that stands without the use of columns to support the roof. Domes work on the principle of gaining maximum strength from minimal use of materials.

After several demonstrations, Fuller patented his dome in 1951. Its first practical use was in 1953, when the Ford Motor Company used a geodesic dome to cover its headquarters' rotunda in Detroit. A conventional dome would have weighed about 160 tons (145 metric tons) and was not technically practical. By contrast, the dome materials on the Ford project weighed only 8.5 tons (7.7 metric tons).

Large and Wind Resistant

In addition to being structurally sound, the geodesic dome offers virtually no resistance to the wind. For this reason, it is used to house sensitive **radar** equipment along the U.S. Defense Early Warning (DEW) line in the Arctic and in other harsh environments. Geodesic domes look somewhat like large golf balls and are capable of withstanding winds exceeding 200 miles per hour (322 kph). Also, because of a geodesic unit's lack of wind resistance, it can be transported by helicopter even under windy conditions.

Geodesic domes can be erected in a very short time. A symphony hall in Honolulu was built in 18 hours and hosted a concert an hour later.

Geodesic domes can stretch over acres of ground without relying on poles or columns for support. This feature makes them ideal for exhibition buildings. A dome was used for the Climatron in St. Louis, Missouri, to house a tropical garden complete with fully grown palm trees. A geodesic

R. Buckminster Fuller poses in front of the 20-story tall geodesic dome that he designed to house the United States's Exhibition at the 1967 World's Fair.

dome also was used to house the United States's Exhibition at the 1967 World's Fair. This dome was 20 stories high.

The geodesic concept is even used in some models of backpackers' tents. The 'round' tents offer more headroom than the conventional A-frame.

Many of Fuller's ideas were discounted as the work of an eccentric, but today his domes can be found throughout the world.

Germ theory

If you have ever suffered from a cold or needed an **inoculation** (shot) in order to go to school, then you know that everyday life exposes you to contagious (easily spread) diseases. Many major health problems are caused by tiny microorganisms such as **bacteria** and **viruses** that are impossible to see

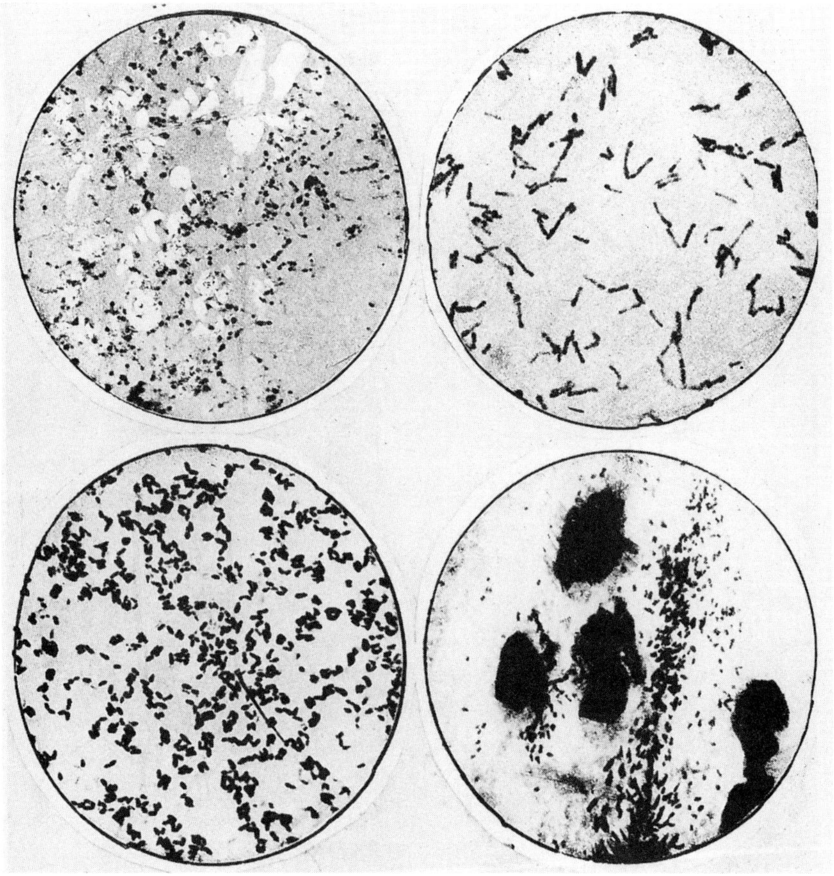

Four kinds of germs: two types of diptheria bacillus (top), pseudo diptheria bacillus (bottom left), and pneumococci bacillus (bottom right). The idea that different microbes such as bacteria or viruses cause a specific disease is called germ theory.

without a powerful microscope. Only in the last 200 years did scientists propose the idea that diseases are caused and spread by microorganisms. The connection between microorganisms and disease is called the germ theory.

Germ theory

In Search of Microorganisms

Early in the nineteenth century, improved microscopes made it possible to see bacteria. But it was not until later in the century that the work of several scientists led to the germ theory—the idea that different microbes such as bacteria or viruses cause a specific disease. Two great scientists that helped shape this theory were Louis Pasteur, a noted French chemist and microbiologist, and Robert Koch, a German doctor.

Pasteur's Research

Louis Pasteur was a serious, hard-working scientist who was trained as a chemist. His experiments with **fermentation** increased his knowledge of bacteria. He applied these insights to disease and eventually developed vaccines against anthrax, a livestock disease, and against rabies. Pasteur also realized that both diseases could be spread by transfer of the bacteria from a diseased creature.

Koch's Photography

Working around the same time as Pasteur was a young German doctor named Robert Koch. Koch had heard of the early work done by Pasteur and was interested in the new germ theory of contagious disease. Koch set up a laboratory in a small room next to his doctor's office and there began experimenting with many types of disease-causing bacteria.

Koch, an amateur photographer, was the first person to take pictures of bacteria through a microscope. In 1876 Koch showed beyond a doubt that a specific kind of bacteria caused the disease anthrax. With his painstaking experiments to support him, Koch published a book in which he announced that specific diseases were caused by specific microbes.

In 1881 Koch worked with tissue taken from an ape that had died of tuberculosis, a lung disease that at the time claimed many lives.

Louis Pasteur's lifelong study of bacteria greatly advanced the development of germ theory.

Before vaccines, many sick people were isolated in quarantine wards of hospitals to prevent their disease from spreading.

Louis Pasteur, French Chemist and Bacteriologist

French chemist and bacteriologist Louis Pasteur was born in Dôle, France, in 1822. His father was a tanner, a craftsman who made leather from animal hides. The young Louis became fascinated by the invisible changes that occurred during the tanning process, changes that turned animal skin into soft, supple leather.

Pasteur attended several prestigious schools, receiving his early training as a chemist. He became professor of physics at the University of Dijon in 1848 and later taught at several universities, including the Sorbonne in Paris.

His first major scientific contribution was in the field of chemistry. His study of acids laid the basis for stereo chemistry, a new branch of chemistry that studied the spatial (space) relationships of atoms. The study of acid continued as Pasteur was asked to find out why some of the greatest burgundy wine produced in France was spoiling.

After observing wine under the microscope, Pasteur noted that wine normally contained yeast cells which were producing the desired alcohol. However, wine that had become sour contained **bacteria**, other microorganisms that produced lactic acid during fermentation and spoiled the wine. Pasteur's experiments showed that fermentation could take place only in the presence of living cells. Pasteur's correct theory of fermentation led him to research in other related fields of microbiology, which included his work in the cause and treatment of numerous diseases caused by microorganisms.

In 1865 Pasteur was asked to find the cause of a disease that was killing silkworms and destroying the French silk industry. Pasteur discovered that microorganisms were responsible for the disease, and devised a method for killing the infected adult silkworms. He was credited with saving the silk industry in France from extinction.

Next, Pasteur experimented with cultures of many disease-causing bacteria. While injecting healthy chickens with the bacteria that cause

→

Koch isolated the rod-shaped bacteria that cause tuberculosis by growing it in culture dishes separate from any other germs. He inoculated healthy guinea pigs with the bacteria and when the guinea pigs became sick, he found the **tuberculosis** bacteria growing in them.

cholera, Pasteur got unexpected results. Because the bacteria sample he used was old, it did not give the disease to the chickens. Instead it gave the animals immunity to the disease. Pasteur had discovered that weakened (or older) microbes make a good vaccine and was successful in developing a similar vaccine for anthrax, a common deadly disease of livestock. Pasteur had discovered the principle of inoculation or vaccination, a disease-preventing shot.

Pasteur also worked on a treatment for rabies, a deadly disease contracted from the bite of an infected rabid animal. He methodically worked on a rabies vaccine using fluid from the spinal cords of infected rabbits and in 1885 saved a young boy from the dread disease in a celebrated case.

Pasteur is equally well-known for his contributions toward milk purification. He once again explored the world of bacteria to determine why milk turned sour. He learned that bacteria were present but could be killed if the milk were heated to above 131° F (55° C). The name "pasteurization" was given to this process.

Pasteur's lifelong study of bacteria greatly advanced the development of germ theory. Once doctors understood the role of invisible germs in the spread of disease, they began to stress hygiene (cleanliness) at home, in the doctor's office, and in schools and hospitals.

The British physician Joseph Lister (1827-1912), a disciple of Pasteur, applied the germ theory in the operating room. Lister's insistence on sterile operating conditions greatly reduced the patient's chance of infection and greatly improved the surgical survival rate.

Pasteur was named director of the Institut Pasteur, a research facility named in his honor in Paris and dedicated to the study of hydrophobia (rabies). He served in this position until his death in 1895.

Koch then removed some bacteria from the guinea pigs and grew it in yet another culture. Next he infected a second group of healthy animals with this cultured bacteria. When those healthy animals contracted tuberculosis, he was sure that the same bacteria was responsible. This long procedure, invented by Koch, became the standard way in which disease-organisms could be identified and is known as Koch's postulates.

Vaccines Become Common

As the 1800s drew to a close, the scientific community worked toward isolating disease-causing bacteria. Most scientists agreed that the organisms that caused many contagious diseases could be isolated and a vaccine for each produced. The scientists who proposed the germ theory contributed much to our knowledge of disease and immunity.

Glaucoma

Glaucoma is the second leading cause of blindness in Americans.

Glaucoma is a serious eye disease that can cause blindness if left untreated. It is caused by an increase in the amount of pressure within the fluid of the eye. The extra pressure can damage the optic nerve that takes messages of sight to the brain. Following **cataracts**, glaucoma is the second leading cause of blindness in the United States.

Glaucoma can be classified into four major groups:

- Congenital or infantile glaucoma occurs during birth or infancy. In many cases the infant develops very large eyes, sometimes called ox eye.
- Open-angle glaucoma, the most common form, is a loss of the field of vision over time. Usually the condition starts with a loss of the peripheral (side) vision so that the person does not realize he or she is losing sight until it is too late for treatment.
- Angle-closure glaucoma occurs when there is a block in the root of the iris, cutting off the flow of fluid within the eye. The resultant increase in pressure causes pain and sudden visual loss. The cornea of the eye appears cloudy and the pupil does not respond to light.
- Secondary glaucoma results from trauma, physical injury, surgery, or other conditions.

Treatment for Glaucoma

To prevent blindness resulting from glaucoma, early treatment is recommended. When someone tests positive for glaucoma, a program of medication is started to keep the eye pressure normal. Another technique currently under development is the use of laser scanners that create three-dimensional maps of the patient's optic nerve. This gives a better picture of the degree of damage to the nerve. Other new medical advances involve using ultrasound to treat glaucoma.

Glycerol

Candy, ice cream, and cake frosting get their creamy consistency from glycerol.

Candy, ice cream, and cake frosting get their creamy consistency from glycerol. Glycerol is a sweet, thick liquid that is classified as an alcohol. The liquid gives toothpaste and facial creams their smooth texture and also prevents tobacco leaves from disintegrating during processing.

Because of its taste and texture, glycerol is used as a sweetener and emulsifier (thickener) in many foods. Glycerol was first found in olive oil in 1783 by Carl Wilhelm Scheele while he was studying a variety of fruit and vegetable materials.

Until the twentieth century, glycerol was obtained almost always as a by-product from the manufacture of soap. Most glycerol is still produced along with soap, but large quantities are now synthesized from **petroleum**. Today, glycerol is used most often to make resins for paints, varnishes, and coatings. Also, cellophane and certain papers become flexible, yet retain their toughness when treated with glycerol. Glycerol is also used to make antifreeze and medicines. Nearly every industry uses glycerol, making it one of the most valuable alcohols.

Glycerol for commercial use is classified according to various grades. These include dynamite grade, yellow distilled glycerin, and chemically pure glycerol.

Gold

The oldest of the metallic elements known to humans is probably gold. The element occurs in a pure form in nature and is widely distributed. Humans might first have seen the element in deposits left by running water, where its brilliant golden color would have stood out from its drab, earthy surroundings.

Man's Relationship With Gold

Gold's chemical symbol, Au, comes from its Latin name aurum, *for "shining dawn."*

Evidence of the use of gold by humans comes from the earliest Egyptian dynasties. Fine gold jewelry shows that Egyptian goldsmiths developed sophisticated techniques for working with the element as early as 2600 B.C. The Bible, too, contains many references to gold, as do Hindu (East Indian) texts going back to at least 5000 B.C..

Gold was so common among early American cultures that it was valued for its beauty rather than its costliness. One Spanish conquistador reported in the fifteenth century that the Incas of Peru treasured copper more highly than they did gold. They used more abundant gold simply for ornamentation and not as a medium of exchange.

The discovery of gold in California in 1848 marked a turning point in American history. Thousands of men and women endured terrible hardships in the great Gold Rush of 1849 to make their fortunes in the California and Colorado foothills.

Gold Has Many Uses

Techniques for working with gold have been known in most cultures for many centuries. For example, the Romans knew how to remove gold from worn-out clothing by treating it with mercury. Christopher Columbus (1451-1506) found that natives in Hispaniola (now Haiti and the Dominican Republic) were able to beat gold into very thin sheets from which they made ceremonial masks. Alchemists (early chemists) of the Middle Ages (A.D. 400-1450) learned how to add gold to glass to make a beautiful gold ruby glass.

Gold is the most malleable (easily shaped) of elements. Pure gold is too soft to be used commercially for most purposes. It is usually alloyed (mixed) with silver or copper for use in jewelry, coins, and decorative objects. Since it is a good conductor of electricity, it is sometimes used in electronic components. Radioactive gold (gold-198) is implanted in tissues as a means of treating some forms of cancer.

Gonorrhea

Gonorrhea is the most common sexually transmitted disease (STD) in the world. It is also the most common bacterial infection in adults. In the

United States, between 2.5 and 3 million cases are reported each year, most occurring in people under age 30. In its early stages, the disease may have no symptoms and therefore can be spread by unsuspecting carriers.

In females, gonorrhea is often without symptoms but can lead to vaginal itching, discharge, uterine bleeding, and other serious complications. In males, gonorrhea causes infection of the urethra (the tube that carries urine) and painful urination.

Each year, almost three million Americans become infected with gonorrhea.

Though not deadly, the disease if untreated can infect other genital organs or the throat. If the infection spreads throughout the bloodstream, it can cause an arthritis-dermatitis (joint-skin) syndrome.

Writers from Biblical times on have described people suffering from gonorrhea. It was not until the late 1800s, however, that scientists began to understand this contagious disease. Today, five types of the gonorrhea or gonococcus organism have been identified.

The first effective treatment for gonorrhea were the sulfa drugs that became available in 1937. During World War II (1939-45), **penicillin** became widely available for the treatment of gonorrhea and other bacterial disease. However, while penicillin and related **antibiotics** are effective in about 90 percent of cases, some strains of the gonococcus are becoming resistant to penicillin.

Gravity

What goes up, must come down. That simple statement describes a very complex force that was a puzzle for millions of years. It took the combined work of many great brains over the centuries to explain gravity.

Scientists still do not know what causes gravity.

In 1613 Danish physicist Isaac Beeckmann suggested the principle of inertia, which describes how difficult it can be to set an object in motion and how hard it is to stop it once it is going. In 1618 he determined that the distance an object falls is related to the square of the amount of time it was falling.

In 1590 **Galileo Galilei** experimented by dropping objects of various sizes from the top of the Leaning Tower of Pisa in Italy. He upset the world of science when he reported that there was no proof that heavier objects fell faster. In 1604 Galileo rediscovered the law of uniform acceleration by rolling balls down an inclined plane.

The great giant of gravitational theory is English physicist **Isaac Newton**. A famous story relates that Newton was sitting in an orchard in the countryside in 1666 and saw an apple fall to the ground. It made him wonder *why* objects fell toward Earth. He theorized that it was because all matter attracts other matter to it.

Newton's general theory of gravitation explained the universal attraction between any two objects. The same force acting on the apple kept the **Moon** in orbit around Earth. Planets, the moons of planets, comets, stars, and galaxies all follow the same laws of gravity.

The first actual measurement of the gravitational force between two objects was performed in 1798 by Henry Cavendish. He suspended a pair of 12-inch (30 cm) lead balls near a pair of 2-inch (5 cm) lead balls, and was able to calculate the force of attraction. His result came within 1 percent of the modern value!

In 1905 **Albert Einstein** published his theory of **relativity**. Einstein's general theory and special theory of relativity allowed scientists to consider the effect of gravity upon something that had been thought to be immune to its influence—light. If light passed close to an object with enough mass, the gravitational pull would force the light to bend from its straight path.

The great giant of gravitational theory is English physicist Isaac Newton.

The greater the mass of an object, the greater its gravitational force. What about an object that has an incredibly high mass, confined to a very small area? The gravitational force would be so great that even light could not move fast enough to escape it. Light would be bent into an orbit around the object. Such massive things are believed to exist. They are called black holes.

The force of gravity has major implications for cosmology (the study of space). If enough invisible "dark matter" can be discovered, its gravitational attraction might be great enough to slow, or even reverse, the expansion of the universe that started with the "big bang" billions of years ago.

It is remarkable to realize that even after 400 years of theories, scientists still do not know what actually causes gravity.

See also **Big bang theory; Galileo Galilei; Newton, Isaac**

Greenhouse effect

For hundreds of years, farmers and florists have relied on greenhouses to grow and ripen vegetables, fruits, and flowers. Greenhouses work because they trap and warm air, and hold in moisture.

Today the greenhouse effect is a phenomenon much discussed among scientists, politicians, and environmentalists around the world. The phenomenon has grown to mean global warming, which is an environmental condition in which damaging gases build up and trap heat within the

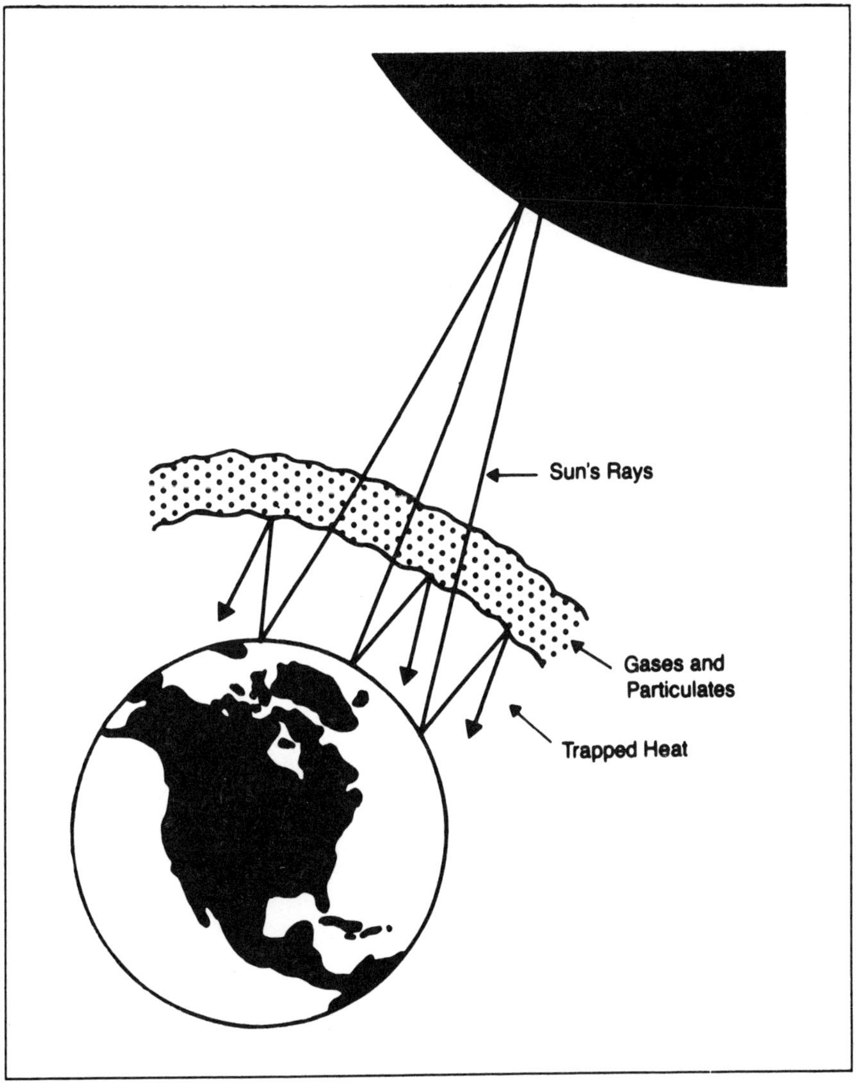

The greenhouse effect, or global warming, is an environmental condition in which damaging gases build up and trap heat within Earth's protective atmospheric shield, the ozone layer.

Greenhouse effect

Earth's protective atmospheric shield, called the **ozone** layer. Greenhouse gases are produced by volcanoes, forest fires, automobile exhaust, farm animals, and power plants. In effect, the theory says that the overabundance of these gases is turning the earth into one giant greenhouse.

The possibility of global warming was first recognized in 1896 by Swedish chemist Svante August Arrhenius. He suggested that the burning of fossil fuels (oil, coal, and gas) would result in a build up of carbon. This could create a greenhouse-like condition that would increase Earth's temperature.

In the 1960s, scientists began to probe a possible link between air pollution and an increase in the Earth's temperature.

Maintaining a Balance

Greenhouse gases are produced by volcanoes, forest fires, automobile exhaust, farm animals, and power plants.

Earth's atmosphere is comprised primarily of **nitrogen**, **oxygen**, and **carbon dioxide.** These three ingredients, along with **methane**, provide Earth with a protective blanket. This blanket regulates how much of the **Sun**'s enormous heat reaches Earth, and how much heat leaves our atmosphere. For this reason, these gases are called greenhouse gases, and their

balance had rarely been questioned until the Industrial Revolution, which began in the mid 1700s.

Through this blanket, the visible wavelengths of radiation from the Sun reach the ground. Once it has hit the surface, this visible light is absorbed and reflected by Earth as infrared radiation that cannot be seen but can be felt as heat. If it were not for the presence of atmospheric gases such as carbon dioxide, water vapor, and other greenhouse gases, the heat would escape out beyond Earth's atmosphere.

A great deal of the carbon dioxide that is released by industrialized societies is absorbed by forests, oceans, and the process of limestone deposition (deposits made through the action of wind, water, etc.). However, these resources are either limited or depleting, and data shows that extra output of carbon dioxide is collecting in the atmosphere. Global warming from the build-up of greenhouse gases can also be made worse by ozone depletion. The thinning or actual puncturing of the ozone layer can allow harmful radiation to reach Earth's surface.

The ozone layer is essential for the protection of life on Earth because it filters out all forms of incoming **ultraviolet radiation** (UV rays) and acts as a protective screen. Exposure to UV-B radiation has been linked to certain types of skin cancer and is also a major cause of **cataracts**.

Technology Contributes

Recently, computers have allowed scientists to develop models that estimate the consequences of global warming while mapping out ways to slow the process. Measurements now show that between 1957 and 1975, the amount of carbon dioxide in the air has increased from 312 to 326 parts per million, a jump of approximately 5 percent. These measurements were collected all around the world from the top of Hawaii's highest mountain, Mauna Loa, to the South Pole, where air was gathered through airplane air-intake systems. It should be noted that this study covers less than 20 years and is inadequate for making long-range projections of ozone damage.

Valuable research continues as scientists try to compare today's atmospheric and air quality with much older historical data. In 1990 two teams of European and American researchers drilled through more than two miles (3.2 km) of ice, searching for trapped air bubbles that would show what the weather was like during the past 200,000 years. Scientists are studying this data to learn how Earth has tolerated different temperature levels and to help prepare us for future climate changes.

The Global Community Responds

As research continues, industries and governments are trying to establish programs that would help protect the environment. The United States banned the use of **chlorofluorocarbons (CFCs)** as an aerosol propellant but not for other applications. Germany has stated that it intends to implement its own ban on CFCs by the year 2000, but other European countries have not committed to such regulations. Besides limiting the use of CFCs, industries worldwide are under pressure to phase out other damaging air pollutants.

An American soldier fires a grenade at the entrenched enemy in Vietnam. Grenades are effective weapons because they allow a soldier to maintain a safer distance when attacking the enemy.

Grenade

Modern grenades are handheld bombs with a time fuse. A soldier activates the grenade by pulling out a safety ring. The grenade is then thrown into

enemy territory, where it explodes. Grenades are effective weapons because they allow a soldier to maintain a safer distance when attacking the enemy. Grenades are also small and easily carried.

The prototype (first version) of the modern grenade appeared in the fourteenth and fifteenth centuries. The first ones were made of bark, glass, or clay pots embedded with large grains of black powder. The explosion was set off by a fuse of powder housed in a quill or a thin tube of rolled metal. Because they looked like pomegranates (an apple-like fruit) with their large seeds, grenades picked up their name from the Spanish word for pomegranate: *granada.* These early grenades were primarily used as incendiary (fire-starting) devices.

Grenades can be constructed to blow into fragments, to penetrate armor, to generate smoke or tear gas, or to fire signal and illuminating flares.

Later grenades were made with round metal bodies that could injure or kill those near the explosion. These metal grenades proved to be dangerous: the fuses were unreliable and the powder occasionally went off before the user could release it. Nevertheless, grenades maintained their popularity with the military. For example, in the 1600s, each infantry company of the British army included five "grenadiers" armed with grenades.

Grenades became more reliable in the twentieth century. They were equipped with firing pins and safety pins. They were launched from rifles and guns as well as thrown by hand. They can be constructed to blow into fragments, to penetrate armor, to generate smoke or tear gas, or to fire signal and illuminating flares.

Gun silencer

A silencer reduces the noise level of a discharged gun by trapping and slowing the release of gases inside the gun's muzzle. In 1908 the Maxim silencer consisted of a cylinder screwed onto the gun barrel. Inside the cylinder were several small chambers, separated from each other by metal rings with holes drilled in the center to allow the bullet to pass through. When the escaping gases rushed into the chambers, they expanded and slowed enough to keep from exploding from the gun's muzzle. As the gases escaped, so did the potential for noise.

During World War II (1939-45), a new variation of the silencer was introduced. Instead of using several chambers separated by metal rings, a simple tube with vent holes was attached to the muzzle of the gun. At the end were several rubber disks through which the bullet passed, and the expanding gases escaped through the holes.

Throughout the years silencers have proved to be less effective when used with revolvers and various types of semi-automatic weapons.

H

Hair care

Curlers and Hair Dryers

A number of nineteenth- and twentieth-century inventions have made hair care and styling easier, more effective, more natural appearing, and longer lasting. In 1866 Hiram Maxim invented the first curling iron. Four years later, two Frenchmen used hot air and heated curling tongs to make deep, long-lasting Marcel waves. Twenty years later, a French hairdresser invented the hair dryer, composed of a bonnet attached to a flexible chimney connected to a gas stove.

Chemicals, heat, and horses' tails have been used to create permanent waves, dyes, and curled hair and to cover baldness.

Also during the late Victorian era, macassar oil from Indonesia came into demand for men's hair. The greasy substance sparked a need for the antimacassar. This was a doily pinned to the back of chairs and sofas to absorb excess oil and protect upholstery.

Permanent Waves

The turn of the twentieth century brought a flurry of inventions and discoveries for hairdressing. At the Paris Exposition of 1900, E. D. Pinaud advertised a brilliantine solution for softening hair, beards, and moustaches and quinine water to control dandruff. In 1906 a German hairdresser working in London applied a borax paste to hair and curled it with a heated iron to produce the first permanent waves. This costly process took 12 hours. Eight years later, Eugene Sutter adapted the method by creating a dryer containing 20 heaters to do the job of waving more efficiently.

Sarah Breedlove became the first African American female millionaire by inventing a method for straightening hair, using a chemical cream and hot combs.

Sutter was followed by Gaston Boudou, who modified Sutter's dryer and invented an automatic roller. By 1920 a Paris beautician perfected a system of curling and drying permed hair for softer, looser curls by using an electric hot-air dryer. This innovation came from the Racine Universal Motor Company of Racine, Wisconsin. **Cortisone**, a medical breakthrough discovered by American Edwin Calvin Kendall at the Mayo Clinic in 1935, replaced earlier coal tar and sulfur preparations to control dandruff.

More types of heaters, permanents, neutralizers to strip away chemicals, and rollers came on the market. The greatest breakthrough came in 1945, when L'Oréal laboratories introduced the first cold permanent wave. This chemical treatment was cheaper and faster than the earlier hot processes. Cosmetologists controlled the amount of curl by varying the diameter of rods used for rolling.

During this same period, frosting—the bleaching of a few prominent strands of hair—came into vogue. In 1959 came the electric curling iron. The next year, a Danish inventor created thermal (heated) hair rollers.

Shampoos and Conditioners

Other products changed hair color and texture. "Shampoo," a word derived from the Hindi word for "massage," dates to 1877. Then English hairdressers boiled soap in soda water and added herbs for health, fragrance, and manageability. In 1905 Sarah Breedlove Walker created a **cosmetics** industry in Indianapolis, Indiana. She became the first African American female millionaire in America by inventing a method for straightening hair, using a chemical cream and hot combs.

By 1927, through the addition of organic materials, hair dyes had become brighter and more natural looking. In 1953 the process of bleaching and dyeing hair was reduced to one step. Another innovation of the period was the addition of a cosmetic base to increase shine. In 1960 L'Oréal laboratories introduced a **polymer** (plastic) hair spray to serve as an invisible net to hold hair in place.

Hair Additions and Replacement

Wigs, hairpieces, and hair extensions, which date to early history, remain popular. Some are expensive handmade versions, while others are cheaper machine-made styles, in which synthetic or natural strands attached to an invisible net or mesh fit snugly against the skull.

One innovation in artificial hair involves surgical implant of a hairpiece. This medical procedure requires suturing (stitching) hair to the scalp.

An alternate method, hair weaving, involves stitching artificial or human hair to existing hair. This process is preferred by many balding males. The intermingling of human and artificial hair produces a natural-looking growth that withstands shampooing, drying and curling, and athletic activity.

Treatment of hair loss is a key area of concern for many men and some women. It received a boost in the 1950s from New York dermatologist Norman Orentriech. His technique was to transplant plugs of four to ten hairs each from the scalp to sections of sparse growth. The placement of plugs requires no stitching, only pressure to assure clotting and proper seating in the new location. Achieving coverage usually involves 10 to 15 once- or twice-weekly sessions in which 20 plugs are moved.

The grafting process, which can be performed in a doctor's office, requires local anesthetic. Despite the pain, expense, and possibility of the hair growing in the wrong direction, the procedure is popular both in Europe and the United States. Louis Feit, a New York plastic surgeon, evolved a more drastic approach to hair replacement by grafting a single plot of tissue containing 350 to 500 hairs. His method, which spurs blood supply to the transplant, improves success rates. Another innovation, the use of antiandrogens, encourages the body to produce hair naturally.

Hallucinogen

Hallucinogens are natural and man-made substances that often cause people to believe they see random colors, patterns, events, and objects that do not exist. Many different types of substances are classified as hallucinogens because of their capacity to produce such hallucinations. These substances come in the form of pills, powders, liquids, gases, and plants that can be eaten.

Street drugs like the hallucinogen LSD are used illegally by those seeking to get "high."

Some users of hallucinogens have reported feeling mystical and insightful. Others are fearful, paranoid (suspicious of others), and hysterical (highly excited). Unlike drugs such as **barbiturates** and amphetamines, hallucinogens are not physically addictive. People do not suffer withdrawal pains when they stop using the substance. However, people can become psychologically dependent upon them. They develop an emotional need for the substance. The real danger of hallucinogens is not their toxicity (poisonousness), but their unpredictability. People have had such varied reactions to these substances that it is virtually impossible to predict the effect a hallucinogen will have on any given individual. LSD (lysergic acid diethylamide) is one such substance whose effect is unpredictable.

Hallucinogenic Plants

Exotic hallucinogenic mushrooms have been used in religious ceremonies since the time of the Aztec civilization (c. A.D. 1300).

Hallucinogens are formed naturally in dozens of plants. These include the peyote cactus, various species of mushrooms, and the bark and seeds of several trees and plants. These natural forms of "psychedelic" (mind-expanding) substances have been available for centuries. In Mexico, exotic hallucinogenic mushrooms called *Psylocybe mexicana,* which contain the fungi *psilocybin* and *psilocin*, have been used in religious ceremonies since the time of the Aztec civilization (c. A.D. 1300). In Europe, the fungus *Amanita muscaria* was thought to have been used by the Vikings (c. A.D. 900). *Amanita muscaria* and its close relative, *Amanita pantherina,* are also found in the United States. Their major hallucinogenic ingredients are ibotenic acid and muscimol.

Today, most people who use hallucinogenic plants in the United States do so illegally. The exception is members of the Native American Church, an organization comprising members of Indian tribes throughout North America. This church advocates the use of mescaline, a form of psychedelic drug found in the peyote cactus. Currently peyote is the only psychedelic agent (substance) that has been authorized by the federal government for limited use during Native American religious ceremonies.

There are a few less-common natural hallucinogens. These include *ololiuqui* (morning glory) seeds, which are ingested by Central and South American Indians both as intoxicants (alcohol-like) and hallucinogens. *Ololiuqui* is used in rituals as a way to communicate with the supernatural. It was first described in the sixteenth century by the Spanish explorer Hernández, who wrote "when the priests wanted to commune with their Gods, they ate *ololiuqui* seeds and a thousand visions and satanic hallucinations appeared to them."

Harmine is another psychedelic agent that has been used for centuries. It is obtained from the seeds of *Peganum harmala,* a plant found in the Middle East. The feeling of exhilaration (happiness) brought about by this drug is sometimes followed by nausea (upset stomach) and sleep. It produces visions similar to those induced by LSD.

The hemp plant provides both marijuana and hashish. They are also considered natural hallucinogens, although their strength is extremely low when compared to others. Marijuana, a green herb from the flower of the hemp plant, is considered a mild hallucinogen. Along with alcohol and nicotine, marijuana is considered one of the gateway drugs whose use can lead to experimentation with stronger drugs. Hashish is marijuana in a stronger, concentrated form.

Man-made Hallucinogens

Even the most potent of these naturally occurring hallucinogens is hardly as powerful and unpredictable as a synthetic (man-made) hallucinogen, such as LSD. LSD became well known in the 1960s when many people sought spiritual experiences through drugs. A form of LSD was first produced in 1938 by Albert Hoffman, a Swiss research chemist. Hoffman accidentally experienced the first "LSD high" when a drop of the material entered his bloodstream through the skin of his fingertip. The effects of LSD are extraordinarily strong. Death from an overdose of hallucinogens is highly unlikely. However, death has resulted from accident or suicide involving people under the influence of LSD.

Halogen lamp

Halogen lamps are used in automobile headlights and spotlights and in floodlights for use in private homes and businesses.

The invention of electric light in the late 1800s dramatically changed life worldwide. Most improvements to the incandescent (glowing) light bulb were made during the early 1900s. By the 1930s, researchers had developed a standard incandescent light bulb that would last for about 1,000 hours. This bulb design has since dominated the market.

Beginning in the early 1970s, as a result of sharply higher world oil prices, scientists became much more concerned with energy efficiency and conservation. Manufacturers of lighting equipment began working to improve the efficiency of incandescent bulbs and extend their useful life. One development was a new type of incandescent light bulb called the tungsten halogen lamp or quartz lamp.

Although halogen lamps cost more than ordinary bulbs, they offer many advantages. They:

- last much longer (about 2,000 to 4,000 hours),
- maintain their light output throughout their entire life, rather than gradually becoming dimmer,
- operate at higher temperatures, allowing them to produce a whiter, brighter light,
- require lower voltage, making them more efficient, and
- use a quartz tube, which can withstand higher temperatures.

Halogen lamps are commonly used in **automobile** headlights, spotlights, and floodlights and are sold for use in restaurants and private homes.

Harpoon

The harpoon is a tool primarily used in the hunting of whales and other sea animals. It was developed by prehistoric people as a variation of the spear.

Originally the harpoon was hand thrown like the spear. Harpoons have three parts: the head, shaft, and line. The double-flued head (like an arrow) pierced the animal's flesh. A line attached to the shaft allowed hunters to retrieve the whale after a chase. This remained the basic design of harpoons for hundreds of years. Its major drawback was that the head tended to slip back out of the entry wound once the animal began to thrash around.

Technological Improvements

Whales have been hunted for their meat for thousands of years by coastal dwellers of Asia and the Americas. In the 1600s, Europeans became interested in whale products. They used whalebones to stiffen ladies' undergarments and whale oil as a fuel. The whaling industry was a mainstay of New England, which built and launched many whaling fleets.

As whaling became more profitable, people sought to improve the effectiveness of the harpoon. In the 1830s, a single-flued model was introduced and became the standard by 1840. Once the single-flued head entered the whale's flesh, the force of the animal pulling against the harpoon caused the head to anchor itself in the whale.

Next, harpoon hunters turned to better methods of propelling their weapons. In the mid-nineteenth century, American ship captain Ebenezer Pierce developed a darting gun that was attached to the shaft of the harpoon. Although the harpoon was still thrust by hand, the darting gun helped immobilize the whale much more quickly than other methods. Upon entering the whale, the barb of the harpoon triggered the gun, which released an explosive-tipped lance. The explosion normally killed or stunned the whale, allowing the whalers to deal with their quarry more easily. Pierce's invention was first used in 1865 and continues to be used by whalers in the Arctic regions.

In 1868 Norwegian whaler Svend Foyn introduced the swivel-mounted harpoon gun. This weapon could be raised, lowered, or rotated on its base. Foyn's redesigned harpoon had a set of hinged barbs that opened once they had penetrated the whale's skin. As the barbs opened, they broke a small vile of sulfuric acid. This lit a fuse connected to a packet of gunpowder. The explosion that resulted had the same effect as Pierce's darting gun, stunning or killing the animal instantly.

Harpoons, circa 1848. The harpoon was developed by prehistoric people as a variation of the spear.

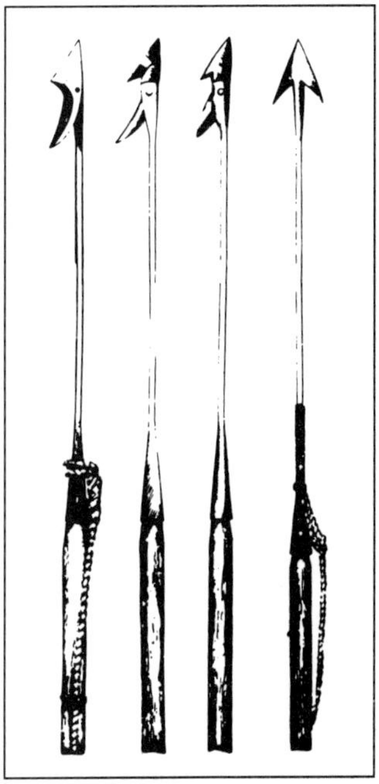

Modern harpoons are still based on modifications of Foyn's design. Fired from a gun mounted on the bow of a whaling ship, the harpoon is composed of a shaft with a line attached, four barbs, and an exploding head at the end. Instead of using a chemical ignition system, however, these harpoons use an electric charge. Once the harpoon has been fired from the boat and is lodged in the whale's flesh, the barbs hold the head in place until it is detonated. The whale is then towed back to the ship by the line attached to the harpoon's shaft.

While many of these hunting methods seem brutal, consider the dilemma of the hunter. For centuries, he hunted alone or in a small band. Their vessel was a small wooden ship tossed about by the high seas. Their prey was a four-ton water creature, difficult to spot and given to sudden dives under water when harpooned. The unlucky hunter who did not cut the harpoon line in time risked major damage to himself and his ship. The drama of the whale hunt has been immortalized in *Moby-Dick* by Herman Melville (1819-1891). Today the United States and many other nations abide by international agreements that govern whale hunting.

Hawking, Stephen William

Physicist and mathematician Stephen Hawking was born in England on January 8, 1942, three hundred years after the birth of his countryman, Sir **Isaac Newton**. In time, Hawking would change our understanding of physics as monumentally as Newton had affected our knowledge of machines and how they work.

Hawking attended Oxford University, where he became very interested in thermodynamics (the science of heat), **quantum mechanics,** and **relativity**. He received his doctorate in physics from Cambridge University. At about this time, he was diagnosed as having amyotrophic lateral sclerosis (Lou Gehrig's disease). This disease causes the central nervous system to degenerate. Today, Hawking's brilliant mind is trapped in an

Stephen Hawking would change our understanding of physics as monumentally as Isaac Newton had affected our knowledge of machines and how they work.

unresponsive body, but he has continued with his work with the help of a special wheelchair and communication system.

Researches Black Holes

Hawking's research has concentrated on the concept of singularity, where an object has extremely high density in a very small volume. Such an object is called a black hole, which is one way a gigantic star can end its life-cycle.

Many consider Stephen Hawking the world's greatest living scientist.

To explain singularity, Hawking has relied on quantum mechanics, which deals with matter at the subatomic (smaller than atoms) level. Hawking has suggested that black holes are not the end result in stellar evolution. The black hole itself continues to evolve by "evaporating," or giving off thermal radiation. If this is the case, the black hole would create "virtual" particles. Virtual particles, unlike "real" particles, cannot be detected. They can only be observed by their effect on other objects. Hawking suggests that when a particle pair is created near a black hole, half disappears into the hole while the other half radiates away as thermal energy.

The rate of evaporation is relative to the mass of the black hole (the smaller the black hole, the faster it evaporates). For mini black holes, the evaporation could occur so fast the result would be an explosion that would leave behind gamma radiation. Hawking is hopeful that someday this theoretical Hawking radiation will actually be detected.

According to Hawking, if **Albert Einstein**'s general theory of relativity is correct, there is another way for black holes to form. At the time of the "**big bang**," when the universe was created from a single piece of matter, there was a great deal of mass in a very small area. Conditions were right for the production of numerous mini black holes following the initial explosion. They could still be in existence throughout the universe today.

In April 1992, NASA announced that its Cosmic Background Explorer satellite had discovered ripples in the fabric of space. Hawking called these remnants of the big bang the "discovery of the century, if not all time."

Hawking has written several best-selling books, including *A Brief History of Time: From the Big Bang to Black Holes,* that have made these difficult concepts understandable to the general public. His current goal is to combine the four basic types of interaction (gravitational, electromagnetic, strong nuclear, and weak nuclear) into a single grand unified theory. The merging of quantum mechanics with the theory of relativity would result in a full quantum theory of **gravity** and greatly advance scientists' understanding of the beginnings of space and time.

One of the largest hearing aids ever made was a huge throne built for King John VI of Portugal in 1819. The hollow carved arms of the chair ended with the wooden mouths of lions. People spoke into these mouths and the sound of their voices was carried by tubes to the king's ear.

Hearing aids and implants

Hearing aids are tools that amplify sound for people who have a hard time hearing. Today many people with hearing impairments can enjoy sounds ranging from a musical symphony to a lively conversation to the rustle of leaves. Millions of hearing aids are sold annually, especially to people over age 65.

A typical hearing aid contains a **microphone** that picks up sounds and changes them into electric signals. The hearing aid's **amplifier** increases the strength of the electric signals. Then the receiver converts the signals back into sound waves that can be heard by the wearer.

The entire mechanism is housed in an ear mold that fits snugly in the ear canal. The power to run the electronic parts is provided by a small **battery**. There are a variety of designs to fit the needs of the wearer, some small enough to be completely concealed by the ear canal.

There are two types of hearing aids, those that conduct sound through the air and those that conduct sound through bone. Often a person who is hearing impaired can use an air-conduction hearing aid. This device amplifies sound and brings it directly to the ear. However, some people have a problem in the transmission of sound through the inner or middle ear. They use a bone-conduction hearing aid, which brings sound waves to the bony part of the head behind the ear and uses the bone to transmit sound waves (or vibrations) to the nerves of the ear.

Other Treatment

Hearing aids cannot be used successfully with some people. One option for them is to have surgery to improve their hearing. The first cochlear implants were done in 1973 and worked to stimulate the remaining undamaged nerves in the inner ear. Today electrical cochlear implants are more sophisticated. They contain speech processors that allow some patients to understand speech without reading lips.

Robert V. Shannon of the House Ear Institute of Los Angeles has developed an auditory brainstem implant for people whose auditory (hearing) nerve has been cut. The implant consists of a tiny microphone, a sound processor, and a transmitter, which are all located outside the ear. An electrode implanted inside the head is connected to the auditory brain stem. When the microphone picks up sound, the device converts sound energy into electric signals that are sent directly to the brain, where they are inter-

Hearing Aids in History

Devices to aid hearing have a long history. The early seventeenth century saw the use of the ear trumpet, which was shaped to gather sound and funnel it into the ear. The Victorian era was known for some of its more elaborate concealed hearing devices, for instance in urns, top-hats, even tiaras. In the 1870s, Alexander Graham Bell began experimenting with the conduction of sound through electrical devices originally intending to help deaf children hear. His experiments led to the invention of the telephone instead, but his work did bring public awareness to the needs of the hearing impaired.

The first electrical hearing aid was made in 1901 by Miller Reese Hutchinson and he called it the Telephone-Transmitter. The first "wearable" hearing aid weighed 2.5 pounds (1.1 kg) and was made by A. Edwin Steven in 1935. During the 1950s, transistors revolutionized electronics and Microtone introduced its compact and powerful transistor hearing aid in 1953.

preted as sound. Although the implant is not enough to restore hearing, it upgrades the level of environmental sound heard by the user and at least one implant volunteer has been able to understand limited human speech.

Heart-lung machine

The heart-lung machine performs all the functions of the heart and lungs when the heart itself is stopped while surgery is performed on it.

Before the heart-lung machine, heart surgeons operated blindly, with the heart still pumping. Their other choices were to slowly chill the patient's body until circulation nearly stopped, or to connect the patient's circulatory system to a second person's system during the operation. These methods were extremely risky. While the idea of a heart-lung machine had been proposed as long ago as 1812, the device was not developed until the 1930s.

The Era of Open-Heart Surgery

In 1931 American surgeon John H. Gibbon, Jr. decided to build a heart-lung machine after a young female patient died of blocked lung circulation. Gibbon, who received his medical degree in 1927 from Jefferson Medical College, began his heart-lung work in 1934 at Massachusetts General Hospital in Boston. In 1946 Gibbon became head of the surgical department at his alma mater and soon secured the backing of Thomas J. Watson, chairman of IBM. With the use of IBM laboratories and engineers, Gibbon's heart-lung machine was perfected.

By 1952 the heart-lung bypass surgery had a 90 percent success rate on animals, and Gibbon decided to use the machine on a human patient. The first attempt failed, although the pump machine worked as required. On May 6, 1953, the second surgery using the heart-lung machine was successfully performed. Despite the deaths of two subsequent patients, the era of open-heart surgery had begun.

Once patients could be kept alive during heart surgery, a whole new range of operations became possible. Congenital (from birth) heart defects could be repaired. Diseased or damaged heart valves could be replaced. Coronary bypass surgery became possible, which meant sewing in a replacement blood vessel to carry blood flow around a blocked section of artery. Thanks to Gibbon's heart-lung machine, open-heart surgery—especially coronary bypass—has become routine throughout the world.

Heat and thermodynamics

Thermodynamics is the study of how heat converts to mechanical energy and vice versa.

In the early days of civilization, heat was a great mystery. Humans considered it an element unto itself, along with earth, air, and water. It was not until scientists such as **Galileo Galilei** (1564-1642) and, eventually, Gabriel Daniel Fahrenheit (1686-1736) constructed the first thermometers that humankind began to truly understand the nature of heat.

The study of heat is called thermal physics. Physics itself is the science that looks at the natural world from its largest components (the universe) to its smallest (atoms). One branch of thermal physics is thermal dynamics. Thermal (hot) dynamics (energy) is the science that studies the relationship of heat and mechanical energy, and how heat converts into energy and energy into heat.

Thermal dynamics really came into its own as a science around the time of the Industrial Revolution (England, 1760-1870). During the Industrial Revolution, people began to rely more and more first on the energy provided by moving water and then on the energy generated by steam.

They used this energy to run the huge machines in the factories that made everything from cloth to furniture to iron and other metals.

These industrialists were interested in making their machines as efficient as possible so they showed great interest in the scientific inquiries of the time. Many of these inquiries focused on heat and how it could be used to drive the engines of machines. What resulted was a better understanding of what heat was and how it worked. This better understanding can be summed up in a set of principles called the laws of thermodynamics.

The first law of thermo-dynamics, also known as the law of conservation of energy, states that energy can be neither created nor destroyed.

First Law of Thermodynamics

The first law of thermodynamics simply states that in a closed system, the total amount of energy is conserved. That is, the energy at the beginning is always equal to the energy at the end. Along the way it may be transformed into a number of different types of energy, such as kinetic (moving energy), potential (stored energy), heat, electricity, etc. However, the total energy at any point is the same. The first law is sometimes known as the law of conservation of energy. Several scientists are given credit for the authorship of this law, which says that energy can be neither created nor destroyed.

Second Law of Thermodynamics

The second law states that, in any system, a certain amount of energy is always transformed into unusable "waste" heat. In a closed system, all of the energy would be eventually converted into heat. This phenomenon is also referred to as entropy.

The law of entropy has been interpreted in many ways by many different sciences. Thermodynamicists claim that it means all systems move inevitably from a state of order to disorder. Environmentalists, on the other hand, use it to prove that the breakdown of the ecosystem can never be reversed—only slowed. Probably the most disturbing application of entropy is the "heat-death" theory of the universe: it states that the universe, being a closed system, contains a finite amount of energy. According to the second law of thermodynamics, that energy can undergo a number of transformations, but it will ultimately be converted into unusable heat. At that time no further transformations can take place—the universe will die, with all matter resting at a temperature a few degrees above **absolute zero**, the point at which all matter stops moving.

Cosmologists (astronomers who focus on the creation of the universe) deny the inevitability of the heat-death theory. They claim that physics as we know it is not universal—that different areas of the universe may observe different laws of physics and that entropy may not apply throughout.

Third Law of Thermodynamics

In its simplest form, the third law of thermodynamics states that if two bodies are at the same temperature as a third body, then they must also be at the same temperature as each other. While this might seem rather simplistic, this law has since been used to prove the inaccessibility of absolute zero. Many experimenters have attempted to reach absolute zero (some approaching it to within one millionth of a degree) but none have succeeded in disproving the third law.

Heating

Humans have searched throughout history for a clean, steady, affordable, and safe source of heat.

When human life began, people grouped around the equator, where it was naturally warm and sunny. As people discovered fire and the furs they got from hunting, they found they could warm themselves artificially. So began the long human trek north into Europe, east into Asia, and south into Africa.

Heat warms human beings in one of three ways—conduction, convection, or radiation.

Conducted heat passes directly from a heated object into the body, as with the heated bricks and warming pans that chased the chill from beds in colonial times in America in the 1600-1700s or electric blankets and battery-operated socks that warm us today.

Convection heats the air around the body. This is the main principle in today's forced-air furnaces.

Radiation resembles the warmth of the sun since heat moves outward in waves and maintains an even temperature. Radiation heat is commonly found in electric coil bathroom wall heaters and quartz space heaters for porches and patios.

Man's Journey Through Time

The most primitive heating systems were cave fireplaces. People improved upon early devices by cutting draft holes to allow smoke to escape. Native Americans used this principle in designing their tepees, which they built around fires and equipped with an opening in the top to channel smoke. Eventually, people built chimneys to direct smoke and soot away from living spaces. However, these open fires required constant tending, removal of ash and creosote, and a screen to protect against falling logs and exploding live coals.

Tile, brick, and iron stoves replaced fireplaces. But they dried out the air and burned anyone who brushed against them. In China, families found greater comfort in sleeping on heated slabs, which they built over hearths. Around 350 B.C., a similar system in Ephesus, Greece, warmed the Great Temple through heated channels in the floor. The first centralized system of heating originated around 100 B.C. with the Roman underground hypocaust. This system directed heated vapors through hollow terra cotta (clay) tubes in walls and floors of homes and public steambaths.

The Dark Ages in Europe (A.D. 450-1000) wiped out Roman innovations, including central heating. Most homes and buildings were drafty

Tile, brick, and iron stoves, such as the one pictured here, replaced fireplaces as the primary sources of heat in homes. But they dried out the air and burned anyone who brushed against them.

and uncomfortable because they were heated with fireplaces and stoves. People stayed warm only when they were near the fire. In 1744 American Benjamin Franklin invented an upgraded version of the stove by regulating the draft and making the temperature easier to control. The Franklin stove remained a staple in most homes until furnaces came into common use.

An improvement on the stove was the creation of the room radiator, which was linked by ducts or pipes to a furnace. In 1831 Jacob Perkins patented a high-pressure model of the heat circulator. This system, which appeared in the United States around 1840, proved clean and dependable. Still, it posed a fire hazard unless the furnace and chimney flue were carefully insulated with fireproof materials and regularly cleaned of ash and soot.

James Watt devised a steam heating system in 1784 in his factory in Birmingham, England. His boiler heated water and directed the resulting steam through pipework and into radiators. Steam heat, however, lacked the efficiency of later methods, which were more easily adapted to the rise or fall of temperatures.

Forced-air or hot-air heating used ductwork to carry heated air from a coal or oil furnace to vents. This method gained popularity in the early nineteenth century, particularly after refinements to fans and pumps were

Solar panels at Rocky Mountain Institute. The cost and size of solar panels and storage cells has limited their popularity.

made. Andrew Franklin Hilyer, a Georgia-born slave who rose to prominence by championing voting rights and business opportunities for blacks, built a fortune on real estate and by creating the heat register and the water evaporator attachment, around 1900.

The hot-air systems contained a built-in problem: they blew dust, fumes, and dirt along with the warmed air. For this reason, electric heating began edging out forced-air heating at the beginning of the twentieth century. Soon radiators were a common sight in homes, businesses, and schools.

Modern Methods

In the 1950s, utilities engineers created a hot water radiant system for baseboard heat. (The baseboard is the wooden board at the bottom of a wall.) A pipe, shielded by a metal cover, followed the baseboard of a room, directing heat to the floor, which was usually the coldest point in a room. This system suited many homeowners because it heated objects rather than the air around them.

The heat pump had been designed by Britain's Lord Kelvin in 1851, but this innovation remained untapped until the twentieth century. The heat pump uses one cycle for heating, then reverses it for air conditioning.

Probably the least used of all current heating systems is the solar collector, which absorbs energy from the sun into tiles, rocks, or water. These things store the energy until it can be circulated or blown over cold rooms. However, the cost and size of solar panels and storage cells has limited their popularity.

★★★ Helicopter

The foundations of the modern helicopter can be seen in the design of the autogiro, sometimes called the gyroplane. The autogiro, developed in 1923 by a Spanish engineer, was an attempt to solve the problem of torque that helicopter designers faced. Torque is the tendency to twist, and that is just what helicopters did. When the power-driven rotor blades swung overhead, the fuselage (body) spun in the opposite direction. Designers responded by creating all sorts of complicated machinery to counteract the torque. The autogiro had a traditional engine at the front to give the craft its power. It added an unpowered, free-rotating rotor tilted up at a slight angle so that air met the rotor blades from below and caused the blades to spin.

While the autogiro was a commercial success throughout the 1920s, the Great Depression in the United States (1929-39) caused sales to drop. Its limited speed, range, and cargo capacity prevented it from being used for military purposes. By the beginning of World War II (1939-45), the autogiro had all but faded from the scene. Its innovative construction, however, paved the way for future helicopter designers.

Modern Helicopter Design

In the 1930s, the prototypes (first versions) of traditional helicopters were being built with varying results. The French had a design in 1936 with two rotors, one above the other, rotating in opposite directions to cancel out torque. This helicopter had set an altitude record of 500 feet (152.5 m) and could reach a speed of 65 miles per hour (104 kph). The German design was similar.

Unlike an airplane, helicopters do not require a long runway to build up speed for liftoff. Instead, helicopters rely on rotating blades to lift them vertically into the air.

At this time Igor Sikorsky made a breakthrough that greatly advanced the helicopter industry. In 1909 he had built helicopters with two rotors, one above the other like the French design, but they were failures. His new idea was to use a small vertical rotor at the end of a long tail boom. This rotor eliminated the complex machinery used in the past to offset the torque prob-

Vertical Flight in History

In the early fourteenth century, the Chinese made toy flying tops that consisted of four rotor blades attached to a spindle. String wound on the spindle was pulled, sending the rotors spinning upward.

In the sixteenth century, Leonardo da Vinci drew sketches of a flying machine with a twisting screw-like wing and an on-board power source. In 1784 two Frenchmen gained widespread attention with a twin-rotor helicopter model operated by a spring-bow mechanism.

Inventors experimented with helicopter designs throughout the nineteenth century but technology lagged. No **engine** was lightweight yet powerful enough to lift a full-size helicopter off the ground. In the early 1900s, with the invention of the gasoline engine, new possibilities opened up for helicopter flight.

In 1907 French inventor Louis Bréguet achieved the first manned helicopter flight, lifting his four-rotor helicopter two feet off the ground for approximately one minute.

lem. This new system, which he first tested in 1940, proved so successful that it is still the most popular design for all types of helicopters. United States military officers were so impressed with Sikorsky's helicopter design that they rushed it into production to be used in World War II.

Several advances in helicopter design took place after the war. The turbine engine solved a major overheating problem that occurred when the helicopter hovered, unable to use moving air to cool the engine. The turbine engine also had more power, an advantage for lifting or rescue operations. Helicopters played a critical role in the Korean and Vietnam wars in the 1950s and 1960s and 1970s and have since become a familiar part of commercial aviation and civilian rescue operations.

Helium

Helium, a gas, is one of the most unusual elements known to science. Helium is far more abundant throughout the rest of the universe than it is here on Earth. In fact, helium was first found to exist on the **Sun**. (Helium was named after the Greek word *helios,* which means "Sun.")

Helium makes up 23 percent of the mass of the universe yet it is a relatively rare element on Earth.

On the Sun, helium is created by the fusion of **hydrogen** atoms, which gives us solar energy. This fusion process, which was first explained by German-American chemist Hans Bethe in the late 1930s, also powers the stars and gives them their light. Scientists estimate that helium accounts for 23 percent of the total mass of the universe. But our atmosphere contains much less helium than could be expected, considering the fact that helium is constantly being produced by uranium and other radioactive substances. Scientists think that most of this helium simply escapes into space because it is so lightweight.

After its discovery in 1895 by Sir William Ramsay, physicists began attempting to liquefy and solidify helium by cooling it. To study the behavior of gases, scientists must measure their characteristics (such as volume, pressure, and temperature) very accurately, and low temperatures make it possible to obtain this critical information. But helium is unique since it remains a gas at temperatures colder than any other element. In fact, it is the only substance known that refuses to freeze solid even at **absolute zero.** The Dutch physicist Heike Kamerlingh Onnes finally succeeded in liquefying helium in 1908, at a temperature only slightly above absolute zero.

Commercial Uses

This early research has since opened up a whole new field of low-temperature science called cryogenics. Today, helium's special properties make it indispensable as a refrigerant in research on **superconductivity** and in the development of modern supercomputers, whose switches must be kept very cold. Also, at very low temperatures, a certain form of helium exhibits a phenomenon called superfluidity—it suddenly transforms into a strange liquid, unlike any other, that has no measurable resistance to flow. This means that it can carry heat hundreds of times more effectively than pure copper, which is the best metallic conductor.

Although helium can be separated from air like other inert gases, the process is relatively expensive. Most commercial supplies of helium are produced from a few natural gas fields in the southwestern United States. Because of its lightness and nonflammability, helium is used for weather balloons and other high-altitude research. Helium is also used to fill children's toy balloons and balloons used for parties.

Although helium's value in high-technology applications will continue to increase, most helium today is consumed in arc welding, where it is used as a gas shield to protect the metal from oxidation. Helium is also used with **neon** in gas lasers and as a carrier gas in chromatographic analysis of chemicals. Deep-sea divers often breathe it in a mixture with oxygen to avoid getting the "bends," which occur when gas dissolved in the blood forms bubbles as the diver rises to the surface.

See also **Krypton; Radon; Xenon**

Hemophilia

The parents of hemophiliacs often try to protect them from rough games or sports, in which even a small injury might result in uncontrolled bleeding.

Few people go through life without suffering an occasional cut, scratch, or bruise. This damage to the skin can result in bleeding if a blood vessel ruptures. The human body does damage control by starting a series of reactions that cause the bleeding to stop. First, platelet cells in the **blood** move toward and attach to the site of the wound. The platelets are held in place by strands of fibrin. The formation of the strands is the key event in what turns into a cascade clotting process. Clotting means the blood thickens to the point where it stops flowing. Without this rapid clotting, people would be in danger of bleeding to death from very minor injuries.

But some people are born with an inherited disease that prevents their blood from clotting. This sometimes fatal disease is called "hemophilia," a word that is made up of Greek and Latin terms referring to "one who loves to hemorrhage or bleed." It was first described by the Islamic (Middle Eastern) surgeon Abu al-Qasim in the tenth century. This genetic disease has directly influenced history. Queen Victoria (1819-1901) of England had several hemophilic sons who died before they had the opportunity to become king. The last Czar of Russia, Nicholas II, was related to Victoria and his son suffered from hemophilia.

As early as the nineteenth century, scientists suspected that hemophilia may be a hereditary disease passed from parents to offspring. They also noticed that generally only males showed the uncontrolled bleeding that is a major symptom of the disease.

What Causes Hemophilia?

Today we know that hemophilia is caused by a small defect in a single human **gene**. When this particular defect occurs, as in classic or type A hemophilia, the body lacks an important protein that helps to form fibrin. This protein is called factor VIII. About 85 percent of all hemophiliacs are missing the gene that instructs the body's **cells** how to produce factor VIII.

Hemophilia B is a less common type of genetic disease caused by a deficiency in another necessary protein, called factor IX. In each type of bleeding disorder, proteins are either missing or deficient and thus fibrin is unable to form. Approximately 2 out of every 10,000 males are afflicted with either type A or type B hemophilia.

Few females suffer from hemophilia. A woman, however, can carry the defective gene and may pass this gene on to her children. Generally,

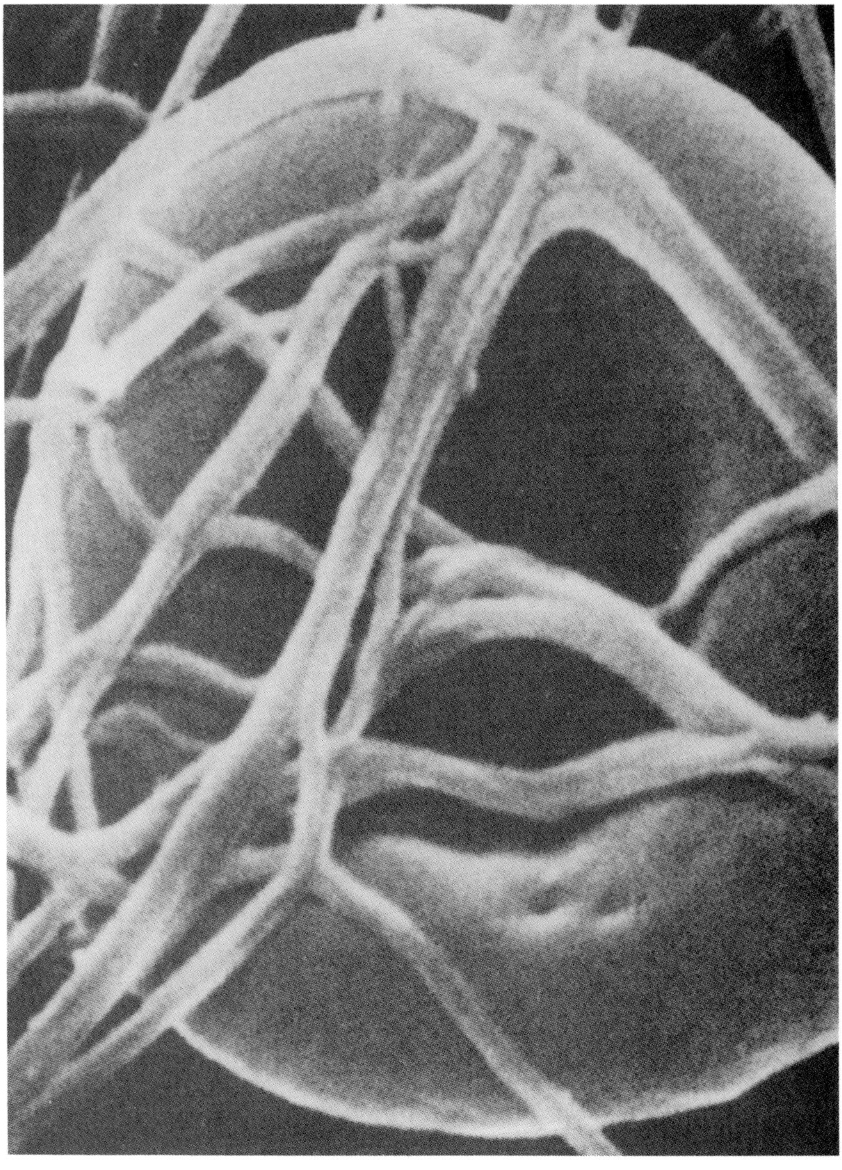

A blood platelet held in place by fibrin. Platelet cells in the blood move toward and attach themselves to the site of a wound.

carrier females will pass their defective gene on to half their daughters, who will be carriers, and to half their sons, who will be hemophilic. For example, Queen Victoria was a carrier, and the genetic profile of her offspring supports this fact.

Treatment

Fortunately, hemophiliacs can be treated with transfusions of concentrated factor VIII protein. This has increased the life expectancy of some hemophiliacs. The protein concentrate can be prepared by combining volumes of blood donated from many humans with normal clotting blood.

The problem, however, is that this method can spread viral diseases that were present in the original donated blood. Many hemophiliacs become infected with these **viruses**, which include the virus that causes **AIDS (Acquired Immune Deficiency Syndrome)** and **hepatitis**. Only recently has a process been invented to detect certain viruses in the blood supply. Cow and pig blood have also been concentrated and used as a therapy, since these animals have higher concentrations of factor VIII than humans. Transfusion reactions and other problems have caused a decline in the use of animal-derived factor VIII, however.

In the early 1980s, genetic engineering eliminated many of the side effects associated with previous factor VIII therapies and has provided hope for many hemophiliacs throughout the world.

Queen Victoria of England had several hemophilic sons who died before they had the opportunity to become king.

See also **Genetically engineered blood-clotting factor**

Hepatitis

Hepatitis is an often fatal disease that causes inflammation (soreness) of the liver. Symptoms include a yellow cast to the skin, malaise (discomfort), liver disease, gastrointestinal (stomach) upset, liver cancer, and scarring of the liver (cirrhosis). When the liver becomes infected and inflamed, it is unable to perform its vital function of helping metabolize food.

Types of Hepatitis

Many types of hepatitis can be prevented. Alcoholic hepatitis is caused by an increase in the fat deposits within the liver. This non-infectious type of hepatitis is common among heavy drinkers. There are also several types of hepatitis caused by **viruses**. These include hepatitis types A, B, C, and D.

In addition to being vaccinated, maintaining clean living conditions, designing adequate sewage facilities, and testing blood supplies for contamination are some methods of helping prevent the transmission of viral hepatitis.

Hepatitis A, also known as epidemic or infectious hepatitis, is spread by the intestinal-oral (stomach-mouth) route. Infection occurs in conditions of poor sanitation and overcrowding. Once the virus has been contracted, it usually takes about 20 to 40 days before symptoms appear. This is known as the incubation period.

Hepatitis B, or serum hepatitis, is transmitted mainly by blood transfusion. Sexual transmission of hepatitis B has also been reported. The incubation period takes between 60 and 180 days. Type B viral hepatitis (HBV) has a higher fatality rate than type A.

Hepatitis C, also referred to as non A, non B hepatitis (NANB), is also transmitted through blood transfusion. Researchers believe it may be caused by several viruses. People infected with hepatitis C usually show relatively mild symptoms that may become ongoing.

In 1977 hepatitis D was discovered. This virus was present only in the liver cells of people who had been exposed to HBV. Scientists strongly believe that this virus requires HBV to survive. Hepatitis D is an important factor in the development of chronic liver disease.

Many other viruses can cause hepatitis, including Epstein-Barr, herpes simplex, measles, **mumps**, and chicken pox.

Preventing Hepatitis

In 1986 a new type of vaccine that used genetic engineering was developed. Scientists were able to make the vaccine by inserting part of the HBV **gene** into baker's **yeast** cells. The yeast cells then produced large amounts of viral protein. This protein resembled the infectious hepatitis virus but lacked certain parts that would cause the disease in humans. The viral protein was then injected into humans as a vaccine. The human body was able to recognize and produce antibodies (disease-fighting cells) against the disease. The final result was a vaccine that allowed the body to produce an artificial but active immunity against hepatitis B. This yeast-derived hepatitis B vaccine is the first such vaccine approved for human use.

Other methods help prevent transmission of viral hepatitis. These include maintaining clean living conditions, designing adequate sewage facilities, and testing blood supplies for contamination.

Heredity

All living things pass on traits from one generation to the next using a systematic set of "blueprints." These blueprints are contained in the long, thread-like **chromosomes** that lie inside the cell nucleus of all living things. On these chromosomes are **genes** which determine the hereditary traits of the offspring.

Egg and sperm cells (or sex cells) are specially formed to carry only one set of the 23 different chromosomes that are normally found in the human body. (Regular body cells have two sets of the 23 chromosomes.) When a mother's egg is fertilized by the father's sperm, the egg inherits one set of chromosomes from each parent for a total of 46 chromosomes.

Some characteristics can only be inherited through genes and chromosomes, including **blood** type, eye color, and sex. These are called hereditary traits. Most characteristics, however, are a result of both heredity and environment. For instance, a person can inherit a general body type, but environmental factors (like diet and exercise) may change that body type.

All living things pass on traits from one generation to the next using a systematic set of genetic "blueprints."

The study of heredity (a science called genetics) began in the 1800s when scientists began trying to explain the existence of different species and variations within the same species.

Lamarck

In the 1800s, French biologist Jean Baptiste de Lamarck believed that acquired characteristics that were routinely used over time would improve. Those characteristics that were not used would simply fade away. Lamarck also maintained that acquired characteristics were inherited from one generation to the next. In other words, Lamarck believed that if a giraffe continuously stretched its neck to reach for food, it would develop a longer neck. And the longer neck would be passed on to the next giraffe generation. Although his belief that acquired characteristics were inherited was incorrect, Lamarck was on the right track. He implied that species undergo long-term evolutionary changes.

Jean Baptiste de Lamarck's studies, though incorrect, pointed to the belief that species undergo long-term evolutionary changes.

Darwin

In 1859 Charles Darwin published his landmark *On the Origin of Species by Means of Natural Selection,* in which he outlined his theory of evolution through natural selection. Darwin believed members of a particular species have slightly different characteristics. In the competition for space, food, and shelter, some of these characteristics make a particular plant or animal better able to survive and produce offspring than others of its species. Therefore, these helpful characteristics would continue to appear in future generations while the less successful ones would disappear as their carriers died out. After many centuries of competition or natural selection, members of a given species might be quite different from their ancestors.

Darwin's natural selection theory stated that changes occur, but it did not explain how the differences in species occurred. Darwin realized he needed to explain the mechanics of variation. He asserted that tiny particles floating in an individual's bloodstream entered the eggs and sperm to determine hereditary characteristics. But a simple blood transfusion experiment between two different types of rabbits proved him wrong. The transfusion did not change the offspring of the rabbits as it should have if Darwin were correct.

Mendel

It was not until 1900 that the second important theory concerning heredity was discovered, although it had been formulated some 45 years earlier. Gregor Mendel, an Austrian monk, had begun experimenting with pea plants at about the same time that Darwin set forth his ideas on natural selection. Through his efforts, Mendel demonstrated that actual physical "hereditary factors" could be transmitted independently. Mendel ultimately established the basic laws of heredity—the missing link to Darwin's natural selection theory—and set the standard for the field of genetics.

Mendel's revolutionary theories, however, were met with disinterest during his lifetime. They remained largely unknown until 1900, when they were rediscovered by Hugo de Vries. De Vries took Mendel's theories further. Unlike the Austrian monk, he believed that rather than arising from gradual or transitional steps, variations occurred in jumps he called **mutations**. This formed the cornerstone of de Vries's mutation theory, which he proposed in 1901.

A mutated fruit fly that has grown antennae in place of legs. Thomas Hunt Morgan's many experiments with the fruit fly helped convince the scientific community that genes were the trait-determiners.

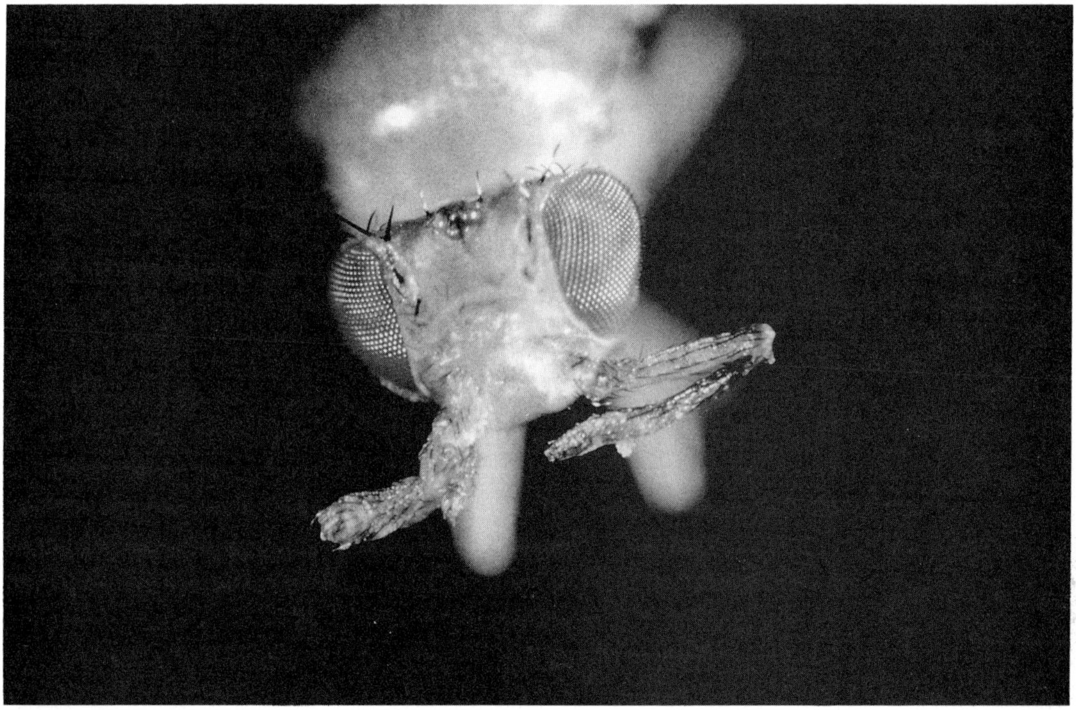

Despite these theories, no biological way to pass on traits had yet been found. Walther Flemming discovered chromosomes during the 1870s but, unaware of Mendel's work, did not understand their genetic significance. In 1903 a young graduate student, Walter S. Sutton, at last made the connection. He observed that during cell division in regular cells, chromosomes were present in pairs. But in the cell division of reproductive (sex) cells, only one member of each pair entered a sperm or egg. The chromosomes became pairs again when the egg joined the sperm in the fertilization process. Sutton saw that this pairing, unpairing, and pairing again paralleled the movement of Mendel's "hereditary units."

By 1909 Wilhelm Johannsen had coined the term "gene" to describe the hereditary units on the chromosome. American geneticist Thomas Hunt Morgan's many experiments with the fruit fly helped convince the scientific community that genes were the trait-determiners and that they are arranged in a certain order on each chromosome. Morgan noticed that all the genes on the same chromosome were usually inherited together. Morgan called these linked genes.

Further experiments showed that traits did not always follow Mendel's basic laws of heredity. Morgan showed that offspring do not

James Watson and Francis Crick helped decipher the genetic code and provided a key to the chemical basis of heredity.

always inherit all of the genes on a chromosome. He called this occurrence crossing over.

Watson and Crick

By 1953 James Watson and Francis Crick developed their spiral model of deoxyribonucleic acid (**DNA**), the building blocks of genes. Their three-dimensional Tinker toy-like model showed where genes occur on the chromosomal strands that make up the DNA. With their model, Watson and Crick helped decipher the genetic code and provided a key to the chemical basis of heredity.

Heroin

The narcotic heroin is derived from morphine.

In the 1890s, Heinrich Dresler (1849-1929) of the Friedrich Bayer pharmaceutical company in Germany found a substitute for **morphine**. At the time, morphine was heavily used as an anesthetic during surgery, but it was unpredictable. Morphine worked because it depressed respiration (breathing) but failed because it sometimes killed severely injured patients. Dresler considered his new drug the ideal morphine substitute. It did not seem to cause physical addiction. It was effective in relieving pain. And it had no depressant effect on breathing.

The Bayer company immediately began marketing the drug under the trade name Heroin, advertising it as, among other things, "the sedative for coughs." Heroin rapidly became popular worldwide.

Four years later it was discovered that heroin, far from being harmless, was one of the world's most addictive substances. Many countries soon passed laws to strictly control its use. Heroin addiction remains a serious problem today because it is a popular illegal street drug. Addicts grow physically dependent on the drug, suffering terrible withdrawal pains if they miss a dose.

High-pressure physics

The study of high-pressure physics includes any type of physical research in which unusually high pressures are used. There is no specific definition for "high pressures," but the term usually refers to pressures that can be produced only with some degree of difficulty. Indeed, much of the history of high-pressure research is the story of efforts to construct instruments by which greater and greater pressures can be produced and by which changes in materials caused by these pressures can be observed and measured.

High-pressure physics helped produce the world's first artificial diamond.

High-pressure physics is of interest in many fields of physics because most physical properties undergo some kind of change with increased pressure. In addition to a decrease in volume, high pressures can cause variations in electrical, magnetic, optical, and chemical properties. For example, some substances undergo phase changes as pressures on them increase. Water is one such substance, capable of existing in seven different phases under various pressures. Electrical conductivity also tends to be a function of pressure. In many metals, conductivity increases approximately 10 percent for every increase of 10,000 atmospheres (units) of pressure applied.

One of the most important practical applications of high-pressure research has been the production of synthetic diamonds. Scientists at the General Electric Company synthesized the first industrial-grade diamonds in 1955, using a device that produced 100,000 atmospheres of pressure. These diamonds are used in cutting machines in factories and jewelry stores.

See also **Electromagnetism**

High-speed flash photography

In 1851 William Henry Fox Talbot produced the first successful high-speed flash exposure during a demonstration for the Royal Institution of London.

High-speed flash photography allows us to see the phases of movement that occur in nature when a bird flies or a cheetah runs. With it, we can see the movements of high-speed machinery and phenomena such as explosions, which normally occur with such great speed that the naked eye cannot absorb the details. These features make high-speed flash photography of great value in scientific and technological studies.

High-speed photography can require exposure times shorter than 1/1,000,000th of a second. Such extremely short exposure times are achieved with the help of magneto-optical shutters. This device uses a very brief electrical current to allow a small pulse of light to pass the shutter and expose the film at a much faster rate. Pulsed light sources, or quick flashes of very intense illumination, further reduce the length of time required to properly expose the film.

Today the movement created by the event being photographed can actually be used to operate the camera's shutter and/or signal the flash. For example, a baseball thrown by a pitcher can break through a light beam to trigger the exposure. This is known as synchronization, and it is intended to accurately trigger the photograph at precisely the correct moment. Advances also have been made in the speeds at which film can be exposed. Today's high-speed flash photography seems limited only by the speed at which film can be mechanically advanced in a camera.

Hologram

The hologram is a true three-dimensional (3-D) image. (The three dimensions are height, width, and depth.) In a hologram, the interference pattern between two beams of coherent light actually distorts the silver material of the film on which the hologram is produced, creating an image with visible depth (see box on p. 550). When viewing it, the eye must re-focus to see foreground and background. One can also look "around" and

A hologram of fruits. Besides creating a fascinating visual effect, the hologram has found a wide range of applications.

"Hologram" comes from the Greek and means "whole picture."

Birth of the Hologram

In 1947 scientist Dennis Gabor was troubled over his electron microscope's inability to show him all the parts of the image he was seeing. Suddenly, the solution came to him. He could take an electron "picture," one that was poor but that contained all the information, and correct it later.

Gabor presented his discovery to his colleagues. They realized that while this process would undoubtedly improve the image, it would require a coherent light source—something that did not exist at that time. (A coherent light source emits all the same length of light waves.) Thus, Gabor's solution, and the word he coined to describe it, "hologram" (meaning "complete picture"), remained merely theoretical for more than a decade.

In 1960 Theodore Maiman introduced the first working **laser**. His invention sent shockwaves through the scientific world and soon came to the attention of two researchers at the University of Michigan, Emmet Leith and Juris Upatnieks, who had been working on **radar**. They saw the laser's coherent light as the final piece in Gabor's puzzle and turned their efforts toward producing the first holographic image. In 1963 they were successful.

Ironically, the hologram never fulfilled its intended purpose, that of improving the resolution of electron microscopes. Still, after the creation of the laser, the hologram has proven an invaluable tool for scientists. For developing the basic principles of holography, Gabor was awarded the Nobel Prize in 1971.

"behind" the subject by tilting and turning one's head. Besides creating a fascinating visual effect, the hologram has found a wide range of applications and has blossomed into a multimillion-dollar industry.

Uses for Holograms

One of the most visible applications of holography is in advertising. Holograms can be found on the covers of magazines, books, and music recordings. In the 1970s, automakers would often show a new car model with a cylindrical hologram. A prospective buyer could walk around the tube and view the vehicle from all angles, though the cylinder was actually empty.

How Holograms Are Created

The hologram is actually a recording of phase interference—that is, the difference between two beams of coherent light. A laser beam is split in two: one beam, called the reference beam, strikes a photographic plate. The second beam, called the object beam, strikes the subject and then bounces onto the plate. The difference between these two beams is the interference caused by the subject, and it is this phase difference that is recorded on the photographic plate. This creates a transmission hologram, which can be seen only in laser light. Another type, called a reflection hologram, can be seen in white light, and it is with this type that most people are familiar.

The medical field was also quick to find a use for holograms. A holographic picture could be taken for research, enabling many doctors to examine a subject in three dimensions. Also, holograms can "jump" mediums—a hologram made using **X-rays** can be later viewed in white light with increased magnification and depth. Holography (the science of holograms) has also been instrumental in the development of acoustical (sound) imaging and is often used in place of X-ray spectroscopy, especially during pregnancies.

A critical application of holography is in computer data storage. Magnetic tape, the most common storage device for home and small-frame computers, is two dimensional, so its storage capacity is limited. Because of its three-dimensional nature, a hologram can store many times more information. Imagine a cube: a two-dimensional cube is a square, offering only one "side" for data to be written upon. A three-dimensional cube offers six "sides," all of which can be used for storage. Optical memories store large amounts of binary data on arrays of small holograms. When viewed by the computer using coherent light, these arrays reveal a 3-D image full of information.

Scientists are now examining the possibility of using holograms to display three-dimensional images, creating true 3-D television and movies.

Credit companies now use holographic symbols on their credit cards. Since the holograms are expensive and difficult to produce, the practice discourages counterfeiting.

See also **Spectroscopy**

Hormone

The body's endocrine system uses hormones to regulate the growth, development, and function of certain **tissues** and also to help turn food into energy (metabolism). Hormones are chemicals produced by the body and released directly into the blood stream.

Hormones are chemicals produced by the body to regulate growth, sex drive, and digestion.

Dozens of different human hormones play important roles in growth, sex and reproduction, **digestion**, blood composition, and stress control. Other animals and plants produce hormones as well. Several hormones may work as a team on the same organ or tissue. Their combined effect may be greater than the sum of their single effects (synergism).

The endocrine system includes the pituitary and pineal glands at the base of the brain, the thyroid and parathyroid glands in the throat, and the adrenal glands, sex glands (ovaries and testes), thymus, and pancreas. The digestive system and other tissues and organs also produce hormones. The body produces a specific amount of a hormone in response to stimuli (signals) from inside and outside. The hormone balance helps keep the body functioning in ordinary or stressful situations—a process called homeostasis.

Identifying the Body's Hormones

The first hormone to actually be isolated and artificially created was the adrenal substance **epinephrine**, by the Japanese-American chemist Jokichi Takemine in 1901. The next milestone was the isolation of the thyroid hormone thyroxine in 1914. Too much or too little thyroxine can cause illness. One of the earliest known thyroid-gland disorders is called Graves' disease, characterized by the thyroid gland's increased size and activity. It was first defined in 1835 by Irish scholar and physician Robert James Graves. Its cause is not known, but Graves' disease is linked to stress and may be hereditary.

In 1921 the Canadian physicians Frederick Banting and Charles Best isolated the pancreatic hormone **insulin**. It was soon used to control diabetes.

Another area of important exploration was with adrenal gland cortex hormones. The cortex (outer covering) of the adrenal glands produces several hormones. The first of these hormones to be discovered was the steroid **cortisone**. Cortisone was the first hormone to be used medically, by the American researcher Philip Hench in the 1940s, to reduce inflammation in rheumatoid arthritis. Other corticoids (adrenal cortex steroid hormones) are hydrocortisone (corticol), corticosterone, and aldosterone.

Cushing's disease, described in 1912 by the American neurosurgeon and physiologist Harvey Cushing, results from excessive production of cortisone, hydrocortisone, and corticosterone. It causes fat redistribution from the lower body to the trunk, facial puffiness, and diabetes.

Hormones produced by the pituitary gland are **human growth hormone** (hGH), or somatotropin (hST). Growth hormone is necessary for body growth and development.

How Hormones Work

No system of the body operates completely independently of the others. The Argentinean physiologist Bernardo Houssay showed how the pituitary gland affected the secretion of insulin and other hormones. For this work, he shared the 1947 Nobel prize in physiology or medicine.

The body must have some way of regulating hormone levels and production. In the 1950s, the British anatomist Geoffrey Harris proposed that a part of the brain called the hypothalamus and the pituitary gland jointly control production of hormones. Feedback or signals such as a hormone's blood level begins the process.

The increase in knowledge about hormones is reflected in studies and treatment of Addison's disease. In 1849 Thomas Addison, a British physician and pathologist (student of disease), matched symptoms of increasing weakness, deeper skin color, and weight loss to an atrophied (dying) condition of the adrenal gland. This was the first medical description of an adrenal gland disease. Addison's disease symptoms are now known to be caused by low levels of two hormones, aldosterone and hydrocortisone, and can be controlled with replacement hormones.

Today scientists are increasing their understanding of hormones' interaction with target tissue cells and how this information can be used in the treatment of illness. Researchers are also interested in the activity of hormone-like substances called growth factors, produced by individual cells throughout the body. Laboratory synthesis and genetic engineering are used to aid this research and to increase the amount of hormones available for treatment.

See also **ACTH (adrenocorticotropic hormone); Endorphin and enkephalin; Genetic engineering; Secretin**

Hula hoop

Think of American fads of the 1950s and what comes to mind? Hula hoops! And oddly enough, they all began in Australia.

Wham-O, the American novelty company, brought us the hula hoop and the Frisbee.

In the 1950s, Australian gym classes used three-foot bamboo rings for calisthenics. In 1957 this form of exercise caught on outside school gymnasiums, becoming a popular form of entertainment. The owners of the American novelty company Wham-O, Richard P. Knerr and Arthur K. "Spud" Melin, heard about the craze and decided to investigate.

Knerr and Melin introduced the hula hoop to neighborhood kids and cocktail party guests in America and discovered that they loved playing with the toy. Immediately Wham-O began production of the American version of the Australian ring. Christened the "hula hoop," the toys were made of vividly colored **polyethylene plastics**. They cost 50 cents to make and sold for $1.98 apiece.

By 1958 the hula hoop was popular internationally. Japan's prime minister Nobusuke Kishi received a hula hoop for his sixty-second birthday. Parisian novelists posed for photographs with them. A Belgian expedition bound for Antarctica brought along 20 of them. German world heavyweight champion Max Schmeling gyrated the hula hoop ringside.

Although their popularity has waned in the decades following their initial craze, hula hoops are still in production and remain stocked in many toy stores around the world.

See also **Frisbee**

Human evolution

For well over a century now, ever since Charles Darwin, Western scientific thought has stated that all of today's species, including man, have arisen from the modification of earlier, simpler forms of life. This theory means that the story of human evolution begins with a creature that most of us today would not consider human.

From Ape to Man

Today's human beings, or *Homo sapiens sapiens,* belong to the Hominid family tree. Hominid means "human types," and describes early creatures who split off from the apes and took to walking upright or on their hind legs. In the overall history of life on Earth, the human species is a very recent product of evolution. There are no human-like fossils older than 4.4 million years, which makes them only one-thousandth the age of life on Earth.

The oldest and one of the first ancestors of all known hominids was thought to be *Australopithecus afarensis,* named for a region in Ethiopia (Africa). What distinguished her from the African pongids (gorillas and chimps) from which she split was that she was clearly built for two-legged walking. She was only about 3 to 4 feet tall and had a brain the size of a chimpanzee. By about 2.5 million years ago, she appears to have evolved into a slightly taller *Australopithecus africanus* with a slightly larger brain. Altogether, there were probably four main species of australopithecines.

From *Australopithecus* came the oldest known hominid to be given the Latin name *Homo,* or "Man." This was *Homo habilis,* called "nimble,"

Darwin's theories sparked considerable religious opposition and public scorn. Clarence Darrow, center, defended a school-teacher fired for teaching the theory in the "Scopes Monkey Trial."

"capable," or "handy" man. He was taller than his predecessors and had a bigger brain. He was the first to use tools made of stone. It is possible that he also was a hunter.

By about 1.5 million years ago, the hominid brain had increased in size to about half of what it is today, and this difference made for a new classification, *Homo erectus,* or "upright man." This was the first hominid to use fire and hand axes and to travel long distances. Early (or archaic) *Homo sapiens* ("wise" or "intelligent") appeared about 300,000 to 400,000 years ago. Although she wore clothing and buried her dead, she still did not have a modern-size brain. It was only about 40,000 years ago that *Homo sapiens sapiens,* or doubly "wise" man, appeared. This creature's body was built exactly like our own and she used real language. With "doubly wise," modern humans turned from strictly hunting and gathering and took to domesticating animals and then plants. Soon settlements turned into real cities and her civilization became based on agriculture.

The Greek philosopher Aristotle was one of the first to speculate on the possibility that organisms had evolved from one another.

Aristotle to Darwin

While the story of human evolution remains a theory (an unproven idea), the course of human evolution seems to have followed this apparently logical, progressive road. However, the discovery and understanding of this road was by no means easy or quick, nor did the discovery story follow the same route. Since about 1850, scientists have had to struggle to try to put together the puzzle of human evolution using the equivalent of randomly found pieces. But well before Charles Darwin offered a unifying theory into which we could try to fit all of these pieces, the foundations of mankind's knowledge about human evolution had been laid.

The Greek philosopher Aristotle was one of the first to speculate on the possibility that organisms had evolved from one another. By medieval times (A.D. 400-1450), such ideas gave way to a literal (exact) interpretation of the first chapter of Genesis in the Bible. This interpretation said that humankind came about from a single, unique act of creation (the Adam and Eve story). During the Renaissance, Italian artist and scientist Leonardo da Vinci (1452-1519) studied evolution. He performed studies that compared humans and beasts and wrote in his notes that, "Man in fact differs from animals only in his specific (characteristics)." However, the literal Biblical interpretation of human origins continued to hold sway, even during the Scientific Revolution of the seventeenth century.

It was during this "Age of Enlightenment" that a Frenchman, Isaac de la Peyrere (1594-1676), discovered what he believed were stone tools made by extremely ancient people. He published his findings in 1655, only to have his book burned. In 1700 the earliest recognition of a fossilized human part was given to a skull fragment found at Canstatt near Stuttgart, Germany. In early eighteenth-century terms, this "ancient" discovery was thought to be about 4,000 years old.

In the overall history of life on Earth, the human species is a very recent product of evolution.

In the same century, however, Swedish botanist Carolus Linnaeus published his *Systema naturae,* a methodical classification of all living things. It was here that he invented the system of binomial nomenclature that is still used today. Nomenclature is Latin for "list of names," and binomial means "two names" in Latin. With this system, the first name designates its genus—a group of organisms that are closely related. Each genus is made up of smaller groups different from each other, called species. This is the second name. Linnaeus classified apes with humans by including them in the genus *Homo,* but gave only modern man the name *Homo sapiens.*

At the same time, French naturalist Georges Buffon endorsed an evolutionary concept of man and the earth itself, arguing that everything in nature develops and changes slowly and continuously.

It was not until the nineteenth century however, that anyone said that modern human beings were the result of an extremely long and slow evolutionary process. In 1809 the French naturalist Jean Baptiste de Lamarck was the first biologist to state boldly that humans evolved from four-footed animals. Although he was wrong when he argued that changes came about because acquired characteristics were passed on to offspring, he was the first real evolutionist.

In 1859, the view of man's history and his place on earth was changed forever by the publication of *On the Origin of Species by Means of Natural Selection,* written by English naturalist Charles Darwin. In this revolutionary book, Darwin stated that all living things achieved their present form through a long period of natural changes. In his 1871 book, *The Descent of Man,* Darwin further argued that man descended from subhuman forms of life.

The Time Line of Discovery

In the 1860s, the physical evidence to back up Darwin's theory began to accumulate. Back in 1856, human fossils had been found in the Neander Valley, near Dusseldorf, Germany. At the time science had no methods for judging their age, and could only say they were very ancient individuals. This particular "ancient" skeleton was the first extinct human form ever recognized, and is today classified as *Homo sapiens neanderthalensis.* Neanderthals were regarded then as being halfway between apes and humans, and it is now known they lived between 100,000 and 70,000 years ago. It is still debated whether Neanderthals evolved into fully modern people or were driven to extinction by an invasion of modern types from elsewhere.

In 1868 five skeletons found in Cro-Magnon caves in southwest France were unquestionably modern or *Homo sapiens sapiens.* They were given the name "Cro-Magnon man." The geological evidence at the site seemed to indicate that they were around 40,000 years old. In 1894 a Dutch paleontologist (bone scientist), Marie-Eugene Dubois, discovered the first known fossil of *Homo erectus,* which came to be called Java man.

After 1900 there was much discussion that a "missing link," or half-man, half-ape, probably existed in the distant past. The first puzzle piece was offered by Charles Dawson, who had turned from his legal profession to archeology. In 1912 Dawson announced a discovery made near Piltdown Common, near Lewes, England. His team had found skull pieces that showed that its owner had a large, modern brain and human teeth set in the jaw of an ape. Although this discovery turned out to be a hoax, many scientists embraced the "Piltdown" man as the missing link. This deliberate fraud confused the human evolution picture for a full 40 years until finally exposed.

Louis Leakey and Mary Leakey dig for bits of prehistoric bone and tools in present-day Tanzania, July 22, 1961.

In 1923 another variety of Dubois's *Homo erectus* was found in China and called Peking man. Nineteen twenty-four became a significant year when Raymond A. Dart, an Australian anthropologist, discovered the first *Australopithecus* fossil in South Africa. Scottish paleontologist Robert Broom supported Dart's theory that it was a primitive ancestor of modern humans, and he found a similar skeleton on his own in South Africa in 1936.

It was not until 1959 that African hominids began to be thought of as older than their Asian cousins. In that year Mary Douglas Leakey found skull fragments at Olduvai Gorge in Tanzania (Africa). The skull proved to be a species of *Australopithecus* called *Zinjanthropus* that was 1,750,000 years old. Her husband, British anthropologist Louis S. B. Leakey, discovered in 1961 what he called *Homo habilis.* The skull of this creature held a larger brain than *Australopithecus* and was between 1.8 and 2 million years old.

In 1974 American archeologist Donald C. Johanson discovered a 4 million-year-old fossil whose scientific name is *Australopithecus afarensis,* but whose popular name became "Lucy." Although her brain was only about one third as large as today's human, the interesting thing about her was that she was completely bipedal (walked upright). Many think that the sudden brain development that later occurred in hominids was the result of having their hands free to use tools.

American archeologist Donald Johanson hold the skull of "Lucy," a 4 million-year-old fossil that he discovered in 1974.

In recent years, new fossil discoveries and genetic evidence have fueled a debate concerning when and where *Homo sapiens sapiens* emerged. In 1988 researchers found numerous fossil fragments in a cave in Israel that suggest that anatomically modern humans lived there about 92,000 years ago. These results suggest that modern humans may have existed much longer than supposed. It also supports the theory that they evolved first in Africa and then spread throughout the world. This notion, called the out-of-Africa model, says that Neanderthals found in Europe and elsewhere were a distinct and parallel human species that came to a dead end.

The out-of-Africa model is opposed by the multiregional model, which argues that modern

Radioactive carbon dating is a scientific technique that has helped pinpoint when certain human ancestors lived.

The Missing Link Found?

In late 1994 fossils of the oldest direct human ancestor were found in Ethiopia. These fossils were remnants of a new species that lived 4.4 million years ago. The discoverers, led by Tim D. White of the University of California at Berkeley, gave the new species the name *Astralopithecus ramidus.*

The bones of 17 chimpanzee-sized individuals were found, with jaws and teeth similar to chimpanzees but other characteristics similar to humans. White's research team is expected to return to the site searching for lower body fossils to determine whether the species walked upright on two legs.

humans arose almost simultaneously and independently in several different places in Africa, Europe, and Asia. As we approach the end of the twentieth century, however, one thing remains constant. The mystery of human origins is a difficult and perhaps even more complicated problem than was ever imagined.

See also **Evolutionary theory**

Human growth hormone (somatotropin)

Human growth hormone is required for normal bodily growth and development—growth of the bone, muscle, and cartilage (connecting tissue) **cells.** Human growth hormone is also called human somatotropin (and abbreviated hGH or hST). The hormone is a protein produced by the pituitary gland, which is located at the base of the skull. Human growth hormone is also necessary for **metabolism** (digestion) of **fat**, water, and minerals.

Lack of normal amounts of growth hormone during childhood can result in a type of dwarfism in which the person is short but of normal proportions and intelligence. Injections of human growth hormone can help the person reach normal or near-normal size if the condition is diagnosed while the bones are still growing.

The bST Controversy

Bovine growth hormone—generally referred to as bST or bovine somatotropin—is being considered for general use by the dairy industry to increase the amount of milk a cow produces. There has been concern that trace amounts of bST in milk will affect human consumers. However, many experts think this is unlikely because bST is active only in cattle and closely related species, such as sheep.

If the body produces too much growth hormone in childhood, the person can have much greater than normal height—what is called gigantism. If the growth hormone overproduction occurs in adulthood, the bones of the feet, hands, and face thicken in a condition called acromegaly.

The pituitary gland releases human growth hormone when it receives a signal from the brain's hypothalamus in the form of growth hormone releasing factor (GHRF). However, release can also be triggered by sleep, exercise, fasting, or hypoglycemia (low **blood** sugar). Release can be inhibited by the hormone somatostatin, as well as lack of sleep, high blood sugar (hyperglycemia), obesity, and a high blood level of free fatty acids.

In 1989 the pharmaceutical firm of Hoffmann-La Roche announced the development of an artificially produced growth hormone releasing factor. When administered, this substancc stimulates the body to produce normal amounts of growth hormone. This may have advantages for a patient because a smaller dose is required than that of growth hormone. Also GHRF can be administered by a skin patch or inhalant, rather than by injection.

See also **Hormone**

Hydrocarbon

Hydrocarbons, containing only **hydrogen** and **carbon**, sound simple, but they can be very complex compounds. Why? Because carbon and hydrogen atoms create a great variety of structures that contain hundreds or even thousands of atoms. Most hydrocarbons come from **natural gas** and **petroleum**. Some are also found in the gases and tars produced by heating coal. Many other hydrocarbons are synthesized, or made artificially.

Hydrocarbons are the basis of products as varied as fabric and industrial feedstock.

Such products as plastic bottles and synthetic fabrics are made of chemicals derived from hydrocarbons.

Our modern way of life depends on hydrocarbon compounds. **Gasoline**, jet fuel, and lubricating oil are mixtures of hydrocarbons that are produced from crude oil. Pipeline gas for heating and cooking is primarily **methane**, the simplest hydrocarbon molecule. Many other products, such as plastic bottles and synthetic fabrics, are made of chemicals derived from hydrocarbons.

Kerosene, butane, and propane are often liquefied and sold in pressurized canisters as a substitute for natural gas. Some hydrocarbon products are widely used as feedstocks (raw materials supplied to a machine or processing plant) for manufacturing petrochemicals—the building blocks for numerous industrial products. The hydrocarbon acetylene is used not only as a fuel in high-temperature welding and cutting, but also as a feedstock for producing other organic chemicals.

Hydrochloric acid

Hydrochloric acid (HCl), one of the strongest of all **acids**, is colorless and very corrosive. On contact, it causes severe burns that should be flushed immediately with water. In air, the acid gives off strong fumes that irritate the throat and lungs if inhaled. Because it does not contain **carbon**, hydrochloric acid is categorized as an inorganic acid. It is often referred to by its traditional name, muriatic acid.

Hydrochloric acid is actually a solution of hydrogen chloride gas in water. A given volume of water can dissolve up to 1,000 times its volume of hydrogen chloride gas. When decomposed, hydrochloric acid produces **hydrogen** and a greenish-colored **chlorine** gas.

Although it is a highly corrosive substance, hydrochloric acid is used by the human body in the digestion of food.

During the Middle Ages (A.D. 400-1450), alchemists began formulating new acids and using them to break down metals and other substances. Around 1625 German chemist Johann Rudolf Glauber discovered a convenient method of manufacturing hydrochloric acid. In one of his earliest experiments, he found that common salt (**sodium** chloride) would produce hydrochloric acid when dissolved in strong sulfuric acid. He also discovered that sodium sulfate acts as a gentle laxative to help cleanse the bowels. Known even today as "Glauber's salt," he called it *sal mirabile* (wonderful salt) and sold it as a cure-all.

In 1824 William Prout discovered small quantities of hydrochloric acid in stomach secretions. This was the first indication that the acid is involved in the human digestive process. Scientists still do not completely understand how the delicate lining of the human stomach protects itself against this corrosive liquid. Hydrochloric acid is involved in digesting our food, and having the right amount of acid in our stomachs is vital to our health. Too much hydrochloric acid can upset the stomach and cause heartburn or even ulcers, while too little acid leads to indigestion.

During the nineteenth and twentieth centuries, hydrochloric acid became an important industrial chemical, which is still used in great quantities today. For example, steel is cleaned in a bath of hydrochloric acid before being galvanized (coated with zinc). The acid is used extensively in metallurgy (separating metals from ores) and food processing, and in the manufacture of synthetic (artificial) rubber and many chemical compounds.

See also **Acid and base**

Hydrofoil

The hydrofoil is a boat-plane that travels the boundary between air and water. The hydrofoil avoids drag by lifting itself out of the water, using wing-shaped skis called hydrofoils that extend into the water from the craft. These hydrofoils function like the wings on a plane, creating lift and flying the hull above the surface of the water. The speed of the hydrofoil keeps the skis skimming over the water as if it were a flat surface.

With a hydrofoil that he built, Alexander Graham Bell set a water-based speed record in 1918 that stood until the 1960s.

The first successful hydrofoil boats were created in the early 1900s. Enrico Forlanini, an Italian airship designer, built a small boat with hydrofoils in 1905. He showed Alexander Graham Bell a later model that impressed the famous American. Bell built one himself, based on Forlanini's patented design, and set a water-based speed record of 71 miles per hour with it in 1918. This record stood until the 1960s.

Although there were small improvements made over the next few decades, hydrofoils did not see commercial use until the 1950s. Then Hans von Schertel, a German scientist, developed his designs for passenger hydrofoils. Italy created their Supramar boats, and Russia and the United States developed hydrofoils with both commercial and military applications.

Hydrofoils today are used by commuter services, fishery patrols, fire fighters, harbor control, water police, and air-sea rescues. For the military, hydrofoils can be excellent small submarine chasers and patrol craft.

Hydrogen

Many scientists believe that hydrogen will someday be much more widely used as a fuel in automobiles and other vehicles.

Hydrogen is the source of most of the energy radiated by the **Sun** and other stars. The enormous amount of heat and light that we get from the Sun comes from hydrogen-fueled reactions. When we burn oil or other fossil fuels, we are actually using the Sun's energy, which was stored millions of years ago by ancient plants.

Hydrogen is remarkable in many other ways as well. Its atom is the lightest, smallest, simplest atomic structure known. It consists of a single **proton** and a single **electron** (these are subatomic particles, the first with a positive charge, the other with a negative electrical charge). Yet these minuscule hydrogen atoms make up 90 percent of all matter in the universe.

On Earth most hydrogen is found combined with other elements in compounds such as water (H_2O). Because hydrogen can combine with all elements except the inert (inactive) gases, it creates more compounds than any other element. By definition, hydrocarbons contain hydrogen, as do almost all other organic (carbon-based) compounds.

The Search for Hydrogen

Before scientists learned the chemical structure of hydrogen and its compounds, people thought that air and water were two of the four basic elements of nature (along with fire and earth). Then, during the seventeenth and eighteenth centuries, chemists began to realize that air contains individual gases, each with different qualities.

The first scientist to collect a gas was Robert Boyle, in the 1650s. Boyle filled a flask with sulfuric acid and dropped **iron** nails into it, turned over the flask and submerged it in more acid. As the iron reacted with the sulfur, bubbles of gas rose upward to collect at the top of the inverted flask. Although Boyle called this gas "factitious air," he had unknowingly isolated (separated out) hydrogen for the first time.

More than 100 years passed before anyone identified hydrogen as a chemical element. In 1766 English chemist Henry Cavendish discovered a non-burning gas that he called "inflammable air," which was hydrogen. Cavendish also produced the gas by dropping pieces of metal into acid. He found that hydrogen is released when zinc, **iron**, or **tin** are dropped into **hydrochloric acid** or dilute sulfuric acid. Other metal-acid combinations did not produce the gas.

Cavendish was not only the first chemist to distinguish hydrogen from ordinary air, but also the first to investigate its properties. He found that hydrogen is extremely light, compared to other gases. In fact, it has the lowest density of any known substance. During the 1780s, scientists were just beginning to experiment with hot-air balloons. When French physicist Jacques-Alexandre-César Charles heard about Cavendish's discovery, he realized immediately that hydrogen would work better than hot air. Charles went on to invent a hydrogen-filled balloon that traveled higher and farther than earlier hot-air balloons.

In one of Cavendish's most important experiments with hydrogen, he discovered that when the gas is combined with air and sparked, the mixture explodes and creates water. Cavendish had proven that water is made of two distinct gases permanently.

The hydrogen atom—with one proton and one electron—is the lightest, smallest, simplest atomic structure known.

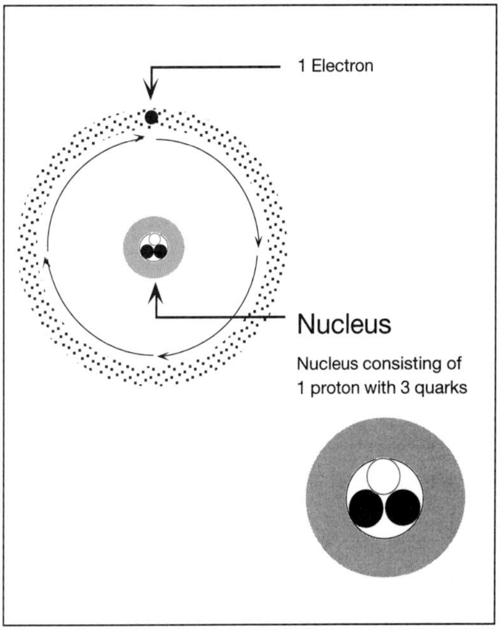

By the nineteenth century, chemists began trying to turn gases into liquids. Normally, pure hydrogen exists as a tasteless, colorless, odorless gas. Many common gases were liquefied simply by applying pressure, while others had to be both cooled and compressed. But hydrogen resisted all such efforts until 1898, when it was successfully liquefied by Scottish chemist James Dewar. Hydrogen became liquid at -423° F (-253° C).

Hydrogen Isotopes

Scientists were aware of only one type of hydrogen atom until 1932. Then American chemist Harold Clayton Urey discovered "heavy" hydrogen (deuterium). This was a second **isotope** (version) of hydrogen and it contained not only a proton and electron, but also a **neutron** (an electrically neutral particle that weighs as much as a proton). Two years later, scientists created hydrogen's third isotope, tritium, which has two neutrons and is radioactive. The thermonuclear fusion of these second and third isotopes is what gives the hydrogen bomb its extraordinary destructive power.

Until recently, scientists doubted that hydrogen could be pressurized into a solid metallic state. If it were practical, solid metallic hydrogen could be formed into pellets of fuel for fusion power generators. Some astronomers believe that metallic hydrogen is a major part of the planets **Jupiter** and **Saturn**. In 1989 American physicists H. K. Mao and R. J. Hemley, using diamond tools, compressed an extremely thin layer of frozen hydrogen into what appeared to be a metal. The transformation took place at a pressure some three million times greater than normal atmospheric pressure.

Henry Cavendish was the first chemist to distinguish hydrogen from ordinary air and to investigate its properties.

Use in Industry

Aside from futuristic uses, hydrogen already has many practical applications in today's industry. By far the largest is for manufacturing ammonia, an important fertilizer and industrial chemical. Margarine, vegetable shortening, and soap are also made with hydrogen, which solidifies oils and fats. Hydrogen is a critical ingredient in certain oil-refining processes, such as the catalytic cracking of petroleum molecules into the lighter **hydrocarbons** found in gasoline. Hydrogen is also used to recover pure metals from their compounds and to create extremely high temperatures for arc welding.

Currently, NASA (National Aeronautics and Space Administration) uses substantial amounts of liquid hydrogen as rocket fuel, and many scientists believe that hydrogen will someday be much more widely used as a fuel in automobiles and other vehicles.

See also **Nuclear fusion**

Hydrogen bomb

The hydrogen (or fusion) bomb is hundreds of times more destructive than the atomic (or fission) bomb.

Even as work on the first **atomic bomb** (fission bomb) was going ahead, some scientists were thinking about an even more powerful weapon, the hydrogen, or fusion, bomb. As far back as the 1920s, scientists explored the possibility that small nuclei might join together—or fuse—to make larger nuclei. In 1938 the German-Austrian physicist Hans Bethe summarized much of this thinking in a theory that explained how stars produced energy.

Creating the Technology for Fusion

Many scientists realized that nuclear fusion could be used as a source of energy on earth, for either military or peacetime applications. But first a method had to be found to carry out and control the reaction. That "if" was a very large one. For one thing, nuclear fusion reactions require temperatures in the range of 72,000,000° F (40,000,000° C). Because of these high temperature requirements, fusion reactions are also referred to as thermonuclear reactions. The challenge of working at such high temperatures was a daunting one.

Atomic Bomb Gets Priority

One scientist, Edward Teller, an émigré from Hungary, began arguing for the development of a fusion weapon, or "super" atomic bomb, during the Manhattan Project. This was the code name for the secret U.S. military program that led to the development of the first atomic bomb during World War II (1939-45). Teller pointed out that such a weapon would be many times more powerful than a fission (atomic) bomb and would provide the United States with an unmatched military superiority.

Teller's ideas were largely ignored, however. The technical requirements for such a bomb were staggering. For example, an important raw material needed for the fusion bomb, tritium (hydrogen-3) had to be made in the lab. Making enough tritium would have taken 80 times the effort currently being spent on the manufacture of plutonium for the fission bomb. There was just not enough time, manpower, or equipment to work on a fusion bomb when the fission bomb had not even been developed.

After World War II ended, many scientists were appalled at the damage done by U.S. atomic bombs dropped on Hiroshima and Nagasaki in Japan. They rejected the idea of building even more powerful weapons. Furthermore, there seemed to be no challenge to the United States's military superiority, based as it was on its possession of fission bombs.

Cold War Heats Up

That situation changed quickly in 1949. In September of that year, the Soviet Union exploded its first fission (atomic) bomb. The following January, President Harry S Truman ordered the U.S. Atomic Energy Commission to begin work on developing a fusion bomb.

The general concept for such a bomb is fairly simple. A fission bomb is surrounded by a large mass of hydrogen. When the fission bomb explodes, it produces temperatures of about 72,000,000° F (40,000,000° C) for a fraction of a second. The fusion bomb goes one step further. The high temperatures are sufficient to initiate a fusion reaction in the hydrogen surrounding the fission bomb.

The greatest technical problem is to find a way to pack hydrogen isotopes together tightly enough to allow fusion to occur and to make the bomb small enough to be transportable. One of the first solutions to this problem was to surround the fission bomb with crystalline lithium hydride, made of lithium-6 and hydrogen-2. When the fission bomb explodes, tritium (hydrogen-3) is produced. This isotope then fuses with hydrogen-2.

The first fusion bomb to be tested by the United States was exploded at Bikini Atoll (island) in the Pacific Ocean on November 1, 1952. It had the destructive power of about seven million tons of TNT, roughly 500 times greater than that of the first fission bombs. The first transportable fusion bomb was exploded by the Soviet Union in 1953. A transportable fusion bomb was not tested by the United States until 1956.

With the break-up of the Soviet Union in 1991 and the end of the cold war, many people in the United States and Russia have urged an end to the buildup of nuclear weapons. Both nations seem willing, but it will take years to dispose of the existing arsenals.

Hydroponics

Early examples of hydroponics, or soil-free agriculture, include the Hanging Gardens of Babylon and the floating gardens of China and Aztec Mexico. Early Egyptian paintings also show the growing of plants in water. The word comes from the Greek and means "water" (*hydro*) and "agriculture" (*geoponics*). Hydroponics is the process of growing plants in liquid nutrients rather than soil.

The Hanging Gardens of Babylon, one of the Seven Wonders of the Ancient World, is an example of hydroponic gardening.

In 1600 the Belgian Jan Baptista van Helmont demonstrated that a willow shoot kept in the same soil for five years with routine watering gained 160 pounds in weight. It had grown into a full-sized plant while the soil in the container lost only two ounces. Clearly, the source of most of the plant's nutrition was from the water, not the soil.

During the 1860s, German scientist Julius von Sachs experimented with growing plants in water-nutrient solutions, which he called nutriculture.

In 1929 W. F. Gericke of the University of California first coined the term "hydroponics." Gericke demonstrated commercial applications for hydroponics and became known for his 25-foot tomato plants.

Hydroponics has been shown to double crop yield over that of regular soil. It can be divided into water culture and gravel culture. Water culture uses the Sachs water-nutrient solution, with the plants being artificially supported at the base. Gravel culture supports the plants with an inert medium such as sand or gravel to support the plants with the water-nutrient solution added.

Hydroponics was used successfully by American troops stationed on non-arable (non-farmable) islands in the Pacific Ocean during the 1940s (World War II, 1939-45). It has also been practiced to produce fresh fruits and vegetables in arid (dry) countries such as Saudi Arabia. In the 1970s, researcher J. Sholto Douglas worked on what he called the Bengal hydroponics system. He sought to simplify the methods and equipment involved in hydroponics so it could be offered as a partial solution for food shortages in India and other developing countries.

Hydroponics, while successfully adopted in certain situations, probably will remain in limited use as long as traditional farming methods in natural soil can support the world's population. However, in the age of space travel, hydroponics may be the natural source for fresh produce.

Hypertension

Blood pressure is the force exerted by blood on the walls of the arteries. When the pressure is too high, the heart is working harder than it should, and blood vessels are being overstressed.

Medical studies show that African Americans suffer from a higher incidence of hypertension than other racial groups.

High blood pressure, or hypertension, can range from mild to severe. In the early 1960s, the results of a long-term study of the residents of Framingham, Massachusetts, convinced doctors that *all* hypertension is dangerous. It can lead to heart attack, stroke, and kidney failure. Hypertension affects 60 million Americans.

Treatment Strategies

Effective drug treatments to control high blood pressure were not developed until after World War II (1939-45). The modern era of drug treatment for hypertension began with the introduction of reserpine in 1953. Reserpine was originally derived from the snakeroot plant, long used by medical practitioners in India, both to lower blood pressure and to induce relaxation. When reserpine was introduced to the market in 1953, it was the first antihypertensive drug to achieve wide clinical use because of its nearly universal effectiveness. It was also widely sold as a **tranquilizer**.

Other drugs took the place of reserpine as manufacturers and researchers focused their attention on antihypertensives, beginning in the 1950s. Thiazide diuretics rid the body of sodium chloride and water by stimulating the flow of urine, and are widely used today to lower blood pressure. Beta blockers are also widely used to treat both angina (heart pain) and high blood pressure. They interfere with the body's signals to increase its heart rate.

Two other antihypertensives were introduced in the 1980s. Calcium channel blockers work by blocking the channel that carries calcium to muscle **cells**. Since calcium is required for contraction of the muscles in artery walls, and it affects the rate at which the heart beats, lowering calcium levels in muscle also lowers blood pressure. ACE (angiotensin converting enzyme) promotes the relaxation of blood vessels.

Drug therapy is not the only way to treat hypertension. From the 1970s on, doctors have recommended exercise, weight loss, eliminating smoking, stress reduction, and a diet low in sodium, caffeine, alcohol, and **cholesterol** as a means of lowering blood pressure.

See also **Blood; Blood pressure measuring device; Neuron theory**

I

Ibuprofen

When ibuprofen was placed on drugstore shelves in 1984, it was the first new over-the-counter (OTC) pain-relief medication to enter the marketplace in a generation. Prior to its introduction, nonprescription pain relief was mainly provided by acetaminophen (introduced in 1955) and aspirin (marketed since 1899).

Today, because of strict monitoring by government agencies, pharmaceutical companies spend millions of dollars researching and testing new drugs before introducing them to the public.

Pharmaceutical Research

Ibuprofen was developed by the Boots Company, a British drug manufacturer and retailer. Early in the 1960s, researchers at Boots identified carboxylic acid as the agent that gave aspirin its ability to reduce inflammation. So the Boots group investigated other carboxylic acids. They soon found one that was twice as strong as aspirin, synthesized it, and tested more than 600 compounds made from these acids. The most effective and useful of these was ibuprofen. Boots began to sell it in 1964 in the United Kingdom as the prescription medication Brufen.

Marketing and Distribution

Ibuprofen appeared in American pharmacies in 1974, when Boots granted a nonexclusive license to the Upjohn Company. Upjohn marketed ibuprofen as the prescription arthritis-relief drug Motrin. A few years later Boots began selling its own prescription-form ibuprofen, called Rufen, in the United States. By 1974 Motrin was one of the most commonly prescribed drugs in the United States.

In the United Kingdom, OTC ibuprofen sales began in 1983. When the U.S. Food and Drug Administration approved OTC sales of ibuprofen at a lower dosage than in prescription form, two major drug companies immediately geared up for a major product-introduction blitz. First the Whitehall Laboratories division of American Home Products came out with Advil. This was soon followed by Nuprin, which was produced by Upjohn and marketed by Bristol-Myers. Both operated under licenses from Boots, which held the worldwide patent for ibuprofen until May 1985.

Soon after, new manufacturers jumped into the lucrative market with products of their own, including Johnson & Johnson's Medipren, Thompson's Ibuprin, and a number of generic and private-label brands. Upjohn and AHP/Whitehall countered with two new ibuprofen products, Haltran and Trendar, promoted as pain relief for menstrual cramps. Sterling Drugs then introduced its own ibuprofen-based menstrual cramps product, Midol 200.

Chemical Content

How does ibuprofen compare to aspirin and acetaminophen? Although the drugs are chemically different, all three give effective relief for minor aches and pains. Ibuprofen causes fewer stomach problems than aspirin, and is more effective for many women in relieving menstrual discomfort. It seems to be more effective for postsurgical dental pain and soft-tissue injuries. But ibuprofen cannot be taken by people with certain conditions, including people allergic to aspirin and women in the third trimester of pregnancy.

Ice-resurfacing machine

The modern ice arena, professional hockey, and touring ice shows all owe their great success to an ungainly machine invented in 1949 by its namesake, Frank J. Zamboni. Born in Eureka, Utah, and raised in Idaho, Zamboni joined his brother George in Paramount, California, at the age of 21. The brothers built a refrigeration plant and began selling blocks of ice to local farmers and householders for their "ice boxes." When the electrical refrigerator began destroying the market for home-delivered ice, the Zambonis decided to build the Iceland Skating Rink across the street from the ice plant.

The rink business did well enough, but Frank Zamboni was bothered by the inefficiency of the nightly cleanup. Five men starting at ten o'clock took as much as an hour and a half to scrape the old ice. The process included cleaning off the scrapings and other debris, squeegeeing up the dirty water, and spreading a fresh layer of water with a hose.

Using a Jeep he had on hand, Zamboni began experimenting in 1942 with ways to mechanize the ice clean-up process. His fourth version, a huge and lumbering contraption completed in 1949, did the job. It scraped the ice, scooped up the debris, squeegeed the surface, and spread fresh water, all in the space of 15 minutes for the entire rink.

The Zamboni machine might have remained a local phenomenon if it had not been for Olympic medalist and Hollywood star Sonja Henie, who ordered two Zambonis to take along on her national tour.

The Zamboni machine might have remained a local phenomenon if it had not been for Olympic medalist and Hollywood star Sonja Henie, who rented practice time at Iceland for her touring troupe. As soon as she saw the ice machine in action, she ordered two Zambonis to take along on her national tour. This advertisement was better than a paid sales force. Ice arena managers nationwide saw the machines and began ordering them, as did the Ice Capades. International exposure came in 1960 when Zambonis were used to clear the ice at the Squaw Valley Winter Olympics in California. Distributorships were soon set up in Switzerland and Japan, and a plant opened in Ontario, Canada.

Today Zamboni machines are used in more than 30 countries. New machines are test-driven down streets from Paramount to Iceland and make a few turns around the rink before shipping. The Zamboni company has no competitors in the United States, though a few exist in Canada and Europe. Zambonis are so common in ice arenas that the term "Zamboni" has become almost generic for all ice-resurfacing machines.

Immune system

The immune system helps the body protect itself against disease.

The immune system is the body's defense mechanism, the means by which the body protects itself against germ invaders. Only in the last century have the parts of that system and the ways in which they work been discovered, and more remains to be clarified.

The true roots of immunology (the study of the immune system and how it functions) date from 1796. Then an English physician, Edward Jenner, discovered a reliable method of smallpox vaccination. He noted that dairy workers who contracted cowpox from milking infected cows became resistant to smallpox. In 1796 Jenner injected a young boy with material from a milkmaid who had an active case of cowpox. After the boy recov-

Turks Fight Smallpox

Since ancient times medical observers had noticed that the body seemed to have powers to protect itself and resist disease. In particular, people who survived some infectious diseases did not suffer from those diseases again during their lifetime. This led to the practice of variolation in Asia, whereby people were injected with a (hopefully) mild case of smallpox to prevent the later development of a severe case of the disease. Lady Mary Wortley Montagu introduced variolation to Britain from the Ottoman Empire (Turkey) in 1720. This was a rather risky procedure, however, because it was always possible for the injected person to develop a serious rather than a mild case of smallpox, and for an epidemic to start.

ered from his own resulting cowpox, Jenner inoculated him with smallpox. The boy was immune. After Jenner published the results of this and other cases in 1798, the practice of Jennerian vaccination spread rapidly.

Germ Theory

It was Louis Pasteur who established the cause of infectious diseases and the medical basis for immunization. First, Pasteur formulated his **germ theory** of disease—the concept that disease is caused by spreadable tiny and invisible microorganisms. In 1880 Pasteur discovered that aged cultures of fowl **cholera** bacteria lost their power to induce disease in chickens. But they still made people immune to the disease when they were injected. American scientists Theobald Smith and Daniel Salmon showed in 1886 that bacteria killed by heat could also confer (give) immunity.

In the late 1890s, antibodies were discovered. Further research showed that each **antibody** acted only against a specific antigen (germ). A new element was introduced into the growing body of immune system knowledge during the 1880s by Russian microbiologist Elie Metchnikoff. He discovered that white blood cells destroyed foreign particles. Up to this point, immune responses were seen as purely beneficial. In 1902, however, Charles Richet and Paul Portier demonstrated hypersensitivity, a condition in which immune responses could cause the body to damage itself. Hypersensitivity to antigens also explained allergies.

Soon knowledge about the immune system's role in rejection of transplanted tissue became extremely important. Peter Medawar's work in the 1940s showed that such rejection was an immune reaction to antigens on the foreign tissue. Donald Calne showed in 1960 that immunosuppressive drugs—drugs that suppress immune responses—reduced transplant rejection, and these drugs were first used on human patients in 1962.

Exciting new discoveries will be made in immunology in the future. Autoimmune reaction—in which the body has an immune response to its own substances—may be one cause of a number of diseases, and research proceeds on that front. Approaches to cancer treatment also involve the immune system. Some researchers speculate that a failure of the immune system may be implicated in cancer. Increasingly sophisticated knowledge about the workings of the immune system holds out the hope of finding an effective method to combat the most serious immune system disorder, **AIDS** (**Acquired Immune Deficiency Syndrome**).

Inoculation

Today, children in industrialized countries routinely receive inoculations against measles, mumps, and whooping cough.

Inoculation is the injection of dead or weakened disease-causing **bacteria** or **viruses** into the human body in order to produce immunity against that disease. The use of inoculation to prevent disease most likely started with smallpox, a dreadful disease first named "variola," in A.D. 570 In the ninth century smallpox was differentiated from measles by a noted Persian physician, Rhazes. Historical accounts mention that epidemics of smallpox raged in many parts of the world.

It is thought that thousands of years ago in China, doctors removed pus and scabs from people who were suffering from smallpox. They put the mixture into scratches made on a healthy person's arm, a procedure later called variolation or scarification. It was hoped that the healthy person would get a mild case of smallpox and would thereafter have immunity against it.

A problem arose some of the time when the inoculated person contracted a serious case of the disease and died. This technique of scratching smallpox fluid onto healthy people was used in China, India, and Turkey for many years. The idea of variolation was introduced to Britain and western Europe in 1717 by an English author, Lady Mary Wortley Montagu. As the wife of the English ambassador to Constantinople (now Istanbul), Lady Montagu had her three-year-old son variolated, as was the practice in Turkey. In the United States Zabdiel Boylston may have been the first to have his son and two slaves inoculated in 1721.

Jenner's Contribution

In eighteenth century Europe, a smallpox epidemic raged. One in ten persons died of the disease, most of them children. Common folk wisdom spread the idea that anyone who contracted cowpox, a similar, milder dis-

A contemporary cartoon that ridiculed Edward Jenner's use of cowpox as a vaccine. Despite the naysayers, thousands of English citizens were vaccinated, including the royal family.

ease of cows, became immune to human smallpox. In 1796 Edward Jenner, an English physician who knew of the practice of variolation, decided to test an idea. He took some cowpox fluid from the sores of a milkmaid named Sarah Nelmes and rubbed it into cuts on the arm of an eight-year-old boy named James Phipps. A few days later James came down with a mild case of cowpox, but soon got over it. Six weeks later, Jenner gave James some fluid from a person who had smallpox. The boy was not affected. He had gained immunity from the inoculation. To describe thc inoculation, Jenner coined the term "vaccine" from *vacca,* Latin for cow, and *vaccinia,* Latin for cowpox.

With the success of his vaccine, Jenner was awarded a sum of money to continue his work. Soon thousands of English citizens were vaccinated, including the royal family. The practice spread to Germany and Russia. Then President of the United States Thomas Jefferson wrote to Jenner congratulating him on his success. When Jenner's vaccination was made available in America, Jefferson insisted that members of his family become vaccinated against smallpox. Jefferson praised Jenner for having found a way to rid humanity of smallpox. For half a century smallpox remained the only disease for which there was a vaccine.

Edward Jenner won praise for having found a way to rid humanity of smallpox.

Attitudes about disease prevention did change slowly during the 1800s. Then people began to realize that filthy living conditions, dirty drinking water, and the lack of proper sewer systems could be associated with outbreaks of disease. Pioneers in health care, including Florence Nightingale, a nurse, and Sir Joseph Lister, a surgeon, brought on the sanitary movement. They called for clean instruments and bedding in hospitals. The practice of scarification became more widespread with the use of a "scarificator," a device with small pointed knives used for bleeding and for injecting smallpox vaccine into the skin.

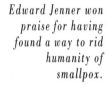

Pasteur's Germ Theory

In the mid-nineteenth century, French chemist and microbiologist Louis Pasteur was developing the **germ theory**. This theory states that specific microorganisms cause specific diseases. During one series of experiments, Pasteur

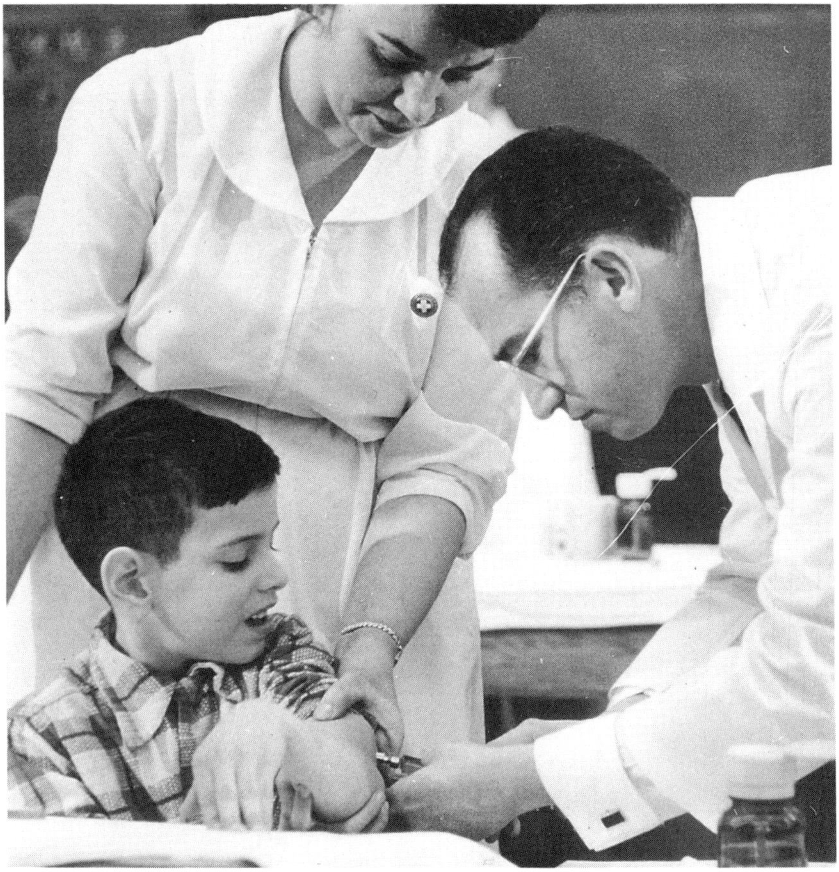

Jonas Salk, the discoverer of the polio vaccine, inoculates a boy with a trial vaccine in 1954. The use of innoculation to prevent disease most likely started with smallpox in A.D. 570.

grew **cholera** bacteria and inoculated healthy chickens with the bacteria. When the chickens failed to get cholera, Pasteur was surprised but later realized that the bacteria had grown old and lost some of its strength. Instead of getting the disease, the chickens received immunity against cholera. This convinced Pasteur that weakened germs would be ideal for a vaccine. They would not be strong enough to cause a serious case of the disease, but would be strong enough to confer immunity.

After a German doctor, Robert Koch, isolated the bacteria that cause anthrax, a deadly disease in livestock, Pasteur set to work to find a suitable vaccine. Later in his career, Pasteur successfully developed a series of inoculations to prevent rabies, using tissue from the brains and spinal cord of rabid rabbits.

Pasteur's assistant, Pierre-Paul-Emile Roux, continued their work in bacteriology, concentrating on the disease diphtheria. He found that not only did a specific bacteria cause diphtheria, but a toxin (poison) produced by the bacteria caused disease symptoms. Once the concept of toxins was established, bacteriologists sought antitoxins to neutralize the harmful effects of toxins as cures for specific diseases.

See also **Germ theory; Immune system**

Instant camera (Polaroid Land camera)

Edwin Land Invents a Fast Camera

Edwin Land revolutionized the photography industries with his instant camera, illustrated below, which has its own processing lab and produces finished prints in less than a minute.

The Polaroid Land camera, now known as the instant camera, is the invention of Edwin Herbert Land. A chemist and physicist, he revolutionized the amateur and professional photography industries with his camera, each of which has its own processing lab and produces finished prints in less than a minute.

The inspiration for his invention came in 1943 during a family vacation, when Land snapped a photograph of his daughter. The three-year-old was impatient to see the results, and even as Land was explaining that

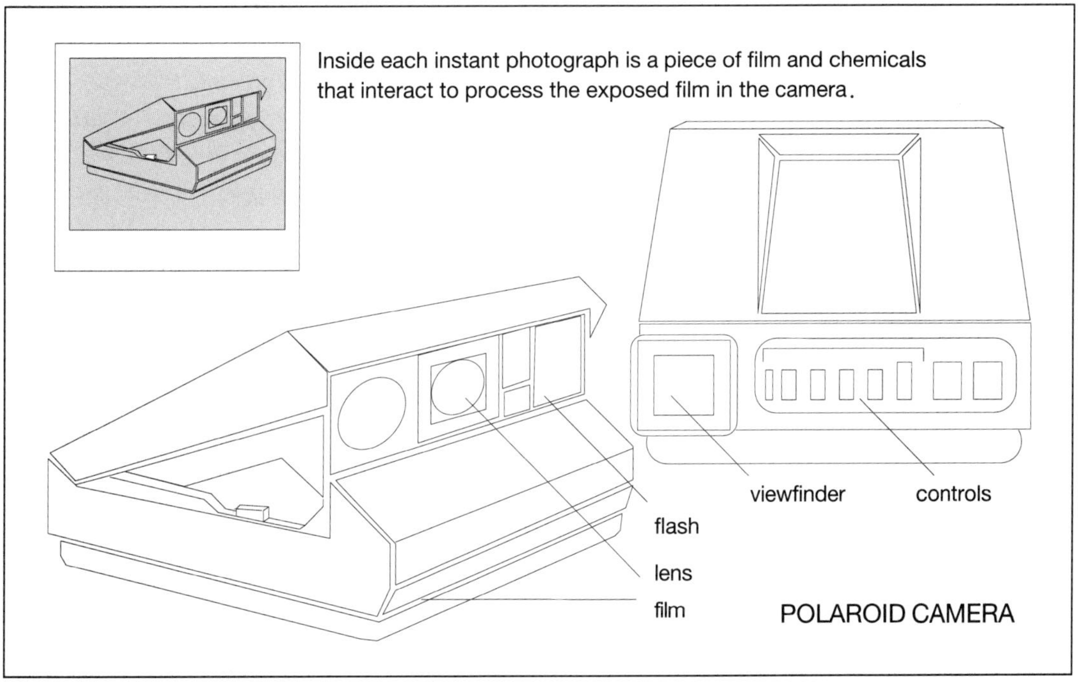
Inside each instant photograph is a piece of film and chemicals that interact to process the exposed film in the camera.
viewfinder
controls
flash
lens
film
POLAROID CAMERA

she would have to wait, he was formulating the first plans for his instant camera.

Usually, once a photo is taken, exposed film must be developed, rinsed, fixed, washed, dried, and printed. The instant camera eliminated several of these steps, delivering a finished print in just minutes. Land's instant camera was a notable success, but he kept improving and producing an even better product. The first photographs from the new camera in 1947 were sepia-toned—a warm, brown color—due to the chemicals involved in the process. But within ten months of the camera's introduction in stores, black and white film was also available. In 1963 the Polaroid Corporation made color film available for its instant camera.

Land produced the instant camera for his three-year-old daughter, who couldn't wait to see her photos.

Polaroid Adds Color

By the 1950s, Polaroid film and cameras had become extremely popular. Amateur photographers liked the Polaroid's immediate results and the chance to "try again" if the results weren't satisfactory. Professional photographers began to use Polaroid backs on studio cameras in order to perform test shots, evaluating lighting, composition, and decor instantly before taking a regular photograph.

The 1972 introduction of the new SX-70 color film meant that the entire development process could be completed within the camera. This eliminated the need to pull the sheet from the camera and peel the negative from the positive print. Processing time was reduced to ten seconds, truly making it an instant process.

Instant coffee

The Early History of Coffee

Coffee was discovered in the ninth century A.D. by an Abyssinian (Ethiopian) goatherd named Kaldi, who noticed that the berries his sheep were eating made them unusually energetic. Kaldi ate one of the berries and felt an exhila-

> **How Instant Coffee Is Made**
>
> Instant coffee is made from regular coffee that has been blended, roasted, and ground. Moisture is then condensed (removed) and it is spray-dried. Freeze-dried instant coffee was developed by the Nestlé company in the 1960s. Freeze-dried coffee involves freezing the coffee extract and placing it in a vacuum chamber, where the moisture is removed. The solid mass that results is processed into granules that can be easily dissolved in water. Freeze-dried coffee is the most popular form of instant coffee because its taste closely resembles regular ground coffee.

rating jolt. He shared his discovery with his fellow goatherds. Later, drowsy monks found that the stimulating berries helped them stay alert during prayers. The Arabs are credited with roasting and brewing the beans to produce a hot drink. The drink reached Europe in the seventeenth century, where it was considered a miracle cure for a variety of ailments.

Today, about one-third of coffee prepared in American households is made from instant coffee.

Efforts to Develop a Tasty Instant Coffee

The first instant coffee, in the form of a liquid extract, was approved by the U.S. Congress in 1838 as a substitute for rum in the rations of American soldiers and sailors. This type of coffee did not catch on with the American public, though. The first powdered instant coffee was developed by a Japanese chemist and was sold to an approving crowd at the 1901 Pan American Exhibition in Buffalo, New York. In 1906 an American chemist improved on the formula and, for the first time, instant coffee was marketed on a large scale.

The production of instant coffee received a boost during World War II (1939-45), when the U.S. War Department purchased the entire U.S. supply for soldiers' rations. Today, about one-third of coffee prepared in American households is made from instant coffee.

Insulin

Insulin is a **hormone** that helps convert carbohydrates into the simple sugar glucose, because glucose is easier for the body to use as fuel. Insulin also regulates the level of glucose in the **blood** and specific events in a cell's

life cycle. Insulin is best known for treatment of the metabolic disease diabetes mellitus. People suffering from diabetes cannot process sugar properly. Without regular shots of insulin, they can slip into a coma and die.

Insulin is produced by the pancreas. It performs its glucose-regulating function with another hormone called glucagon. Together the two hormones ensure that the body stores and uses the correct level of glucose to meet its energy needs.

Insulin's other roles in body metabolism include:

- helping muscle and **fat** cells take in and use glucose,
- inducing production of glycogen,
- encouraging storage of fat within cells by preventing its use as a fuel, and
- stimulating the movement of **amino acids** into cells so they can produce protein in various phases of the cell cycle.

In 1982 insulin became available as a genetically engineered product called Humulin that is identical with human insulin. Humulin is produced jointly by Genentech and Eli Lilly and Co.

Interferon

Diseases fall into two categories: those caused by **bacteria** and those caused by **viruses**. Great strides have been made in fighting bacterial diseases with the use of antibiotic drugs. Less progress has been made in developing effective anti-viral drugs.

Since the 1950s, scientists have been trying to learn how the body protects itself against viruses. The human **immune system** fights off bacteria, but what attacks viruses?

Early in his studies of the viral disease influenza (or flu), Scottish doctor Alick Isaacs became interested in something he called the viral interference phenomenon. He found that this interference seemed to be caused by something inside the cell. In 1957, while working with the visiting Swiss scientist Jean Lindenmann, Isaacs found that chick embryos injected with influenza virus released very small amounts of a protein that destroyed the virus and also slowed the growth of any other viruses in the embryos. Isaacs and Lindenmann named the interfering protein "interferon."

Opposite page: Discoverers of insulin Frederick Banting and Charles Best with the first dog to be kept alive with the hormone, 1921.

Further research showed that interferon was produced within hours of a viral invasion (whereas antibodies take several days to form), and that most living things, including plants, can make the protective protein. Interferon was seen as the cell's first line of defense against viral infections, and its discovery was expected to pave the way for successful treatment of viral diseases. However, researchers soon found that interferon is species-specific. Only human interferon will work in human beings. Also, the body produces it in only tiny amounts. These difficulties slowed interferon research.

Interest in interferon was revived in the late 1960s when Ian Gresser, an American researcher in Paris, discovered that interferon stopped or slowed the growth of tumors in mice and also stimulated the production of tumor-killing lymphocytes. Gresser and Finnish virologist Kari Cantell then developed a way to make interferon in useful amounts from human blood cells. Monoclonal antibodies, first produced in 1975, made large-scale purification of interferon possible. The mid-1980s saw the advent of genetically engineered interferon. Research into interferon's ability to kill cancer cells is now very active and interferon has been used successfully against leukemia and osteogenic sarcoma, a bone cancer. The use of interferon to treat viral diseases such as rabies, **hepatitis**, and herpes infections also continues.

Internal combustion engine

By 1770 the **steam engine** had been developed to the point that the French engineer Nicolas Cugnot used one to successfully propel a three-wheeled vehicle. Steam-power reigned supreme in industry for nearly a century, eventually giving way to the internal combustion engine as an accepted and common source of power.

French Invention

In 1824 French physicist Nicholas Carnot (1796-1832) published a book that set out the principles of an internal combustion engine that would use an inflammable mixture of gas vapor and air. Basing his work on Carnot's principles, another Frenchman named Jean-Joseph-Éttien Lenoir presented the world with its first workable internal combustion engine in 1859. Lenoir's motor was a two-cycle, one-cylinder engine with slide valves and used illuminating gas (coal gas) as a fuel. It also used an electrical charge, supplied by a **battery**, to ignite the gas after it was drawn into the cylinder. Lenoir sold several hundred of his engines, and he adapted his engine to power a carriage. As a result, he is credited with inventing the first documented gas-powered **automobile**.

Lenoir's primitive two-stroke engine was inefficient, however. In 1862 another Frenchman, Alphonse Beau de Rochas, designed a four-stroke engine that would overcome many problems associated with the gas engines of that time. The four-stroke engine doubles the motion of the piston required to accomplish intake, compression, and exhaust, and by doing so greatly increases the efficiency of the engine.

German Refinements

These early French developments and theoretical suggestions were combined by two Germans, Nikolaus Otto and Eugen Langen. Their first workable four-stroke internal combustion engine was patented in 1876 and was tremendously successful; over 50,000 engines were sold in the 17 years following its introduction.

The internal combustion engine was the result of the struggle to develop an engine that would capture a much greater amount of the potential energy of its fuel than the steam engine did.

The "Otto-cycle" four-stroke engine made practical the development of modern **automobiles**, **aircraft**, motorcycles, and other vehicles.

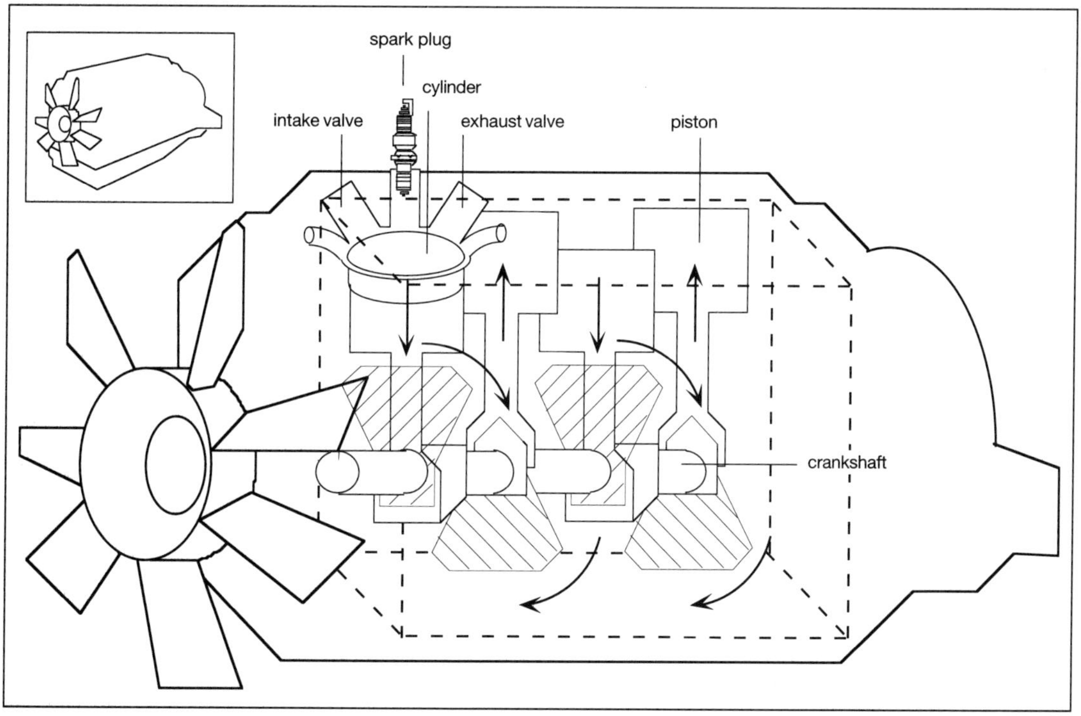

Internal combustion engine

Gottlieb Daimler, who once worked for Otto, made several refinements to Otto's engine. He devised a cam system so that engines with more than one cylinder could be ganged together. Daimler's two-cylinder gas engine was built in a "V" shape, and he is credited with this innovation.

Although supporters of Carl Benz and others dispute his claim, Daimler is also credited with developing the first internal combustion engine to burn **gasoline**—not coal gas—as its fuel. This development was made possible by the invention of the carburetor by Benz's partner, Wilhelm Maybach.

The internal combustion piston engine has been developed into various configurations:

- "in-line" engines with 2, 4, 5, 6, or 8 cylinders,
- the "V" shape engines of 2, 4, 6, 8 or more cylinders,
- engines with two opposing pistons for each combustion chamber (used in Henry Ford's first car),
- multi-cylinder opposed piston engines, and

A worker secures Cadillac engines during final assembly at the General Motors plant in Detroit. Gottlieb Daimler is credited with developing the first internal combustion engine to burn gasoline as its fuel.

Steam Engines

Steam engines use the heat of combustion to boil water, converting it into steam vapor. By this action, the water volume is expanded greatly, creating pressure. The pressurized steam that results is then transferred to an engine, where it is used to push a piston. A crankshaft is used to convert the reciprocating (up and down) motion of the moving piston into rotary (turning in a circle) power that can be used to turn lathes or other machine tools, or, as was the case in Nicolas Cugnot's use, to propel a vehicle.

The greatest drawback of the steam engine is its inefficiency. Much of the potential energy of the burning fuel is converted to heat, and more is lost in the transfer of the steam to the engine. Early in the Industrial Revolution (England, 1760-1870), inventors struggled to develop an engine that would capture a much greater amount of the potential energy of the fuel. This was the impetus behind the development of the internal combustion engine.

- engines—usually of an odd number of cylinders—arranged radially (used extensively in airplanes).

In addition to piston-driven, gas-powered internal combustion engines, other internal combustion engines have been developed, such as the Wankel engine and the gas turbine engine. **Jet engines** and diesel engines are also powered by internal combustion.

See also **Engine**

In vitro fertilization

The term *in vitro* comes from the Latin for "in glass." "In vitro fertilization" (IVF) is the term used to describe the fertilization of an egg in a glass petri dish in order to produce a baby. The procedure is used to treat infertility, a condition in which a woman's egg cannot be fertilized while it is in her womb. An egg is removed from the womb, is fertilized with sperm, and is returned to the womb. Although IVF was done more than 100 years ago with rabbits, the term today usually refers to fertilization of a human egg.

The first time in vitro fertilization was successfully carried out on a human was in England in 1978. An English couple underwent the procedure and produced Louise Brown, the first human baby conceived outside the womb. Since then, more than 3,000 babies conceived in this way have been born.

An English couple underwent in vitro fertilization and produced their baby, Louise Brown, in 1978.

IVF has become a widely used method of infertility treatment, offered by hundreds of medical centers around the world. Since the mid-1980s, cryopreservation—freezing—of eggs (or embryos) has become common, to have them for future use if the current IVF attempt does not result in pregnancy. Legal and moral questions have resulted, dramatized by the 1984 plane crash death of a husband and wife who left behind cyropreserved embryos in Australia. Who has the right to dispose of cyropreserved embryos? How long can they be stored? Who owns them? Is an embryo an heir of its parents?

Variations on IVF

Several variations on IVF are now in use. One is gamete intrafallopian transfer (GIFT), in which eggs and sperm are gathered and prepared as in IVF but are placed into the woman's fallopian tube, uniting there rather than in the petri dish. Another is zygote intrafallopian transfer (ZIFT) in which one or more zygotes—fertilized eggs before they start to divide—are transferred to the fallopian tubes.

Ionosphere

The ionosphere is a region of Earth's upper atmosphere. It is distinguished from all other regions by its layers of electrically charged particles, which can reflect radio waves transmitted from Earth's surface.

Opposite page: Composition of Earth's atmosphere. The ionosphere is a region of Earth's upper atmosphere. Its layers of electrically charged particles distinguish it from all other regions.

The ionospheric layers are identified by letters. Layer D is between 50 and 60 miles (80 and 100 km). With the loss of the **Sun**'s light and radiation at night, layer D nearly disappears, forcing some radio stations to broadcast during daytime hours only.

Layer E is located between 60 and 90 miles (100 and 150 km) and is referred to as the Kennelly-Heaviside layer. The two F layers, F1 and F2,

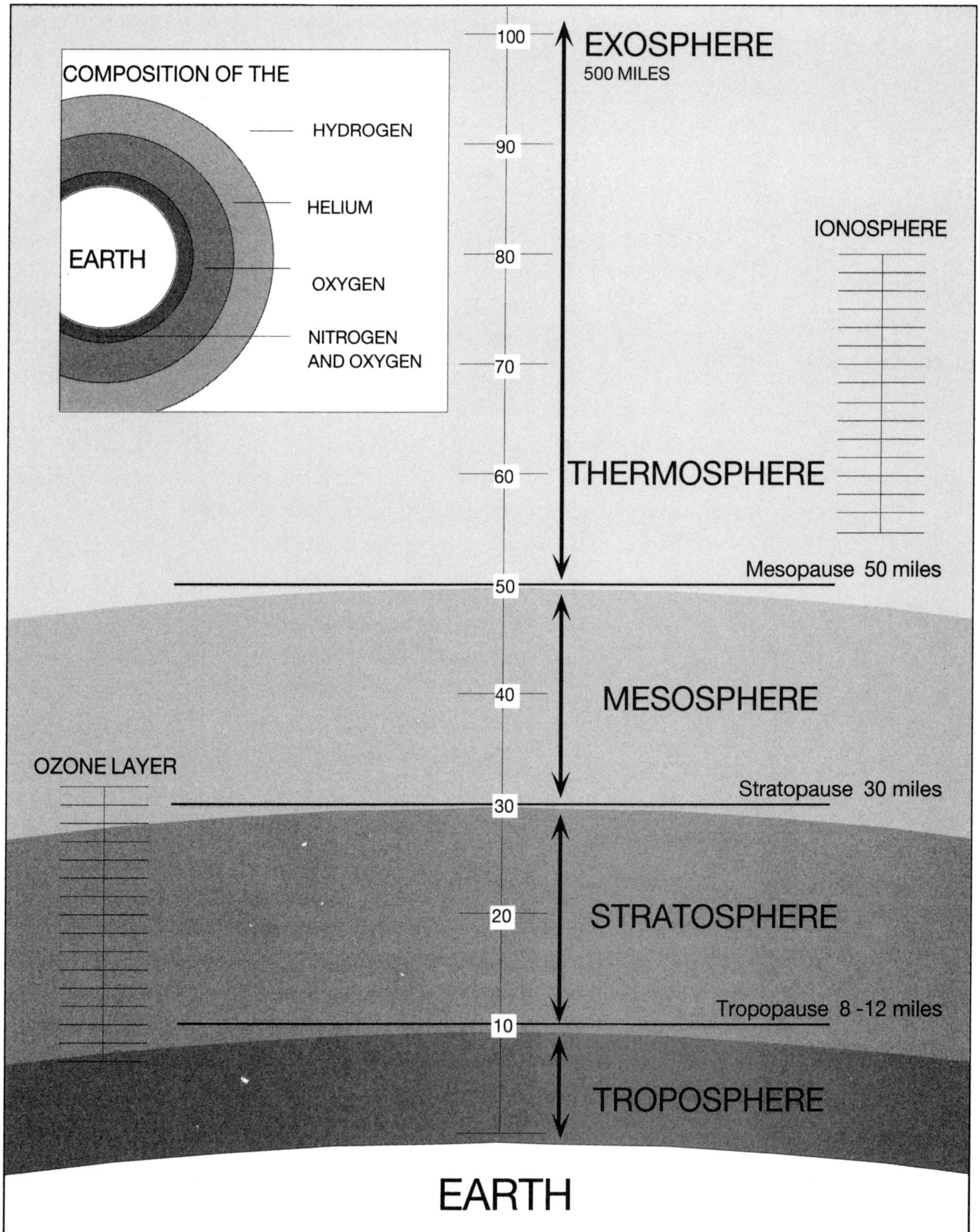

The Earth is 7,926.41 miles in diameter at the equator.

are located between 90 and 190 miles (150 and 300 km) upward and are referred to as the Appleton layers. These upper layers rise at night and extend the reception of radio signals. This is why radio stations from distant locations can be picked up clearly during the night.

The electrically charged particles in the ionosphere receive radio waves from one source and bounce them back to another.

Transmitting Radio Waves

The ionosphere was used for radio transmissions before it was known to exist. In 1888 Heinrich Rudolph Hertz made the first radio transmissions. In 1901 Guglielmo Marconi signaled the letter "S" from Cornwall, England, to St. John's, Newfoundland, Canada. He proved that signals could be transmitted beyond the curve of Earth's surface but failed to explain how it was done. Until then it was thought that radio waves could only be transmitted in a straight line and over only a short distance from any given point. Marconi's transmission created a flurry of debate. The theory of a reflective layer in the upper atmosphere was born. This theory suggested that radio waves bounce off the ionosphere and return to Earth.

The one individual who contributed the most to current knowledge of the ionosphere was Edward Appleton. He became interested in radio science while a signal officer during World War I (1914-18). In 1924 Appleton became a professor at King's College, at the University of London. It was here that he and graduate student Miles Barnett experimented with the fading in and out of radio signals. They were able to calculate the distance to the upper atmosphere layer and discovered its daily fluctuations. Appleton discovered in 1926 that reflections were being received off layers that were higher and lower than the known layer. These were the D and F layers. Alexander Watson-Watt, one of Appleton's research associates, was the first to call the reflective layers the "ionosphere."

Makeup of the Ionosphere

Additional information about the ionosphere has been gathered through the use of rockets and satellites. For example, we know the region is very hot, with temperatures estimated at 1200° K. Sunspot activity and solar flares are directly responsible for disturbances in the radio-reflective layers and can completely disrupt communications.

A combination of ultraviolet radiation from the Sun and cosmic rays (highly energized particles) from space are responsible for the activity in the ionosphere. The particles become electrically charged and separate negatively charged electrons out of **oxygen** and **nitrogen** molecules.

As ionized particles travel through the atmosphere, they cause the formation of the aurora borealis and aurora australis (northern and southern lights) at altitudes of 80 to 300 kilometers over latitudes 20° to 25° north and south. At these latitudes, Earth's magnetic field is vertical to the planet's surface.

As research continues, scientists are learning about the ionosphere's cycles as well as its causes and effects. Computer modeling is making it easier to predict changes so radio frequency adjustments can be made automatically.

See also **Atmospheric composition and structure**

Iron

Iron is the fourth most common metal in Earth's crust and forms the basis for much of modern industry. Iron is also a very common element among the other heavenly bodies in the **solar system**, including the **Sun**, planets, and meteors. Its atomic symbol is Fe, from the Latin word *ferrum.*

The Iron Age lasted more than 5,000 years, from prehistoric times until the end of the Industrial Revolution.

Iron in the Blood

The metal iron is also present in animal life. It appears with **oxygen** in the **blood** of vertebrates, including human beings. Iron oxide forms the substance that gives red blood cells their color and carries oxygen to the tissues of the body. (In invertebrate species [animals without backbones], iron is only one of many possible minerals to serve this function. Many spiders and insects, for instance, have green blood that contains copper.) Iron is also an essential agent in the transporting of oxygen in respiratory (breathing) **enzymes**.

Primary sources of iron in the diet are meats, especially liver, and nuts and legumes (beans and peas). Discovery of this dietary element was made by American pathologist George Whipple in 1925 during his research on the human liver.

Iron in Industry

As a metal, iron has been in use since before recorded history. Around 4000 B.C. the first iron tools and weapons were made from iron from meteorites that had crash landed in the Middle East. The Iron Age eventually spread across Europe and into China. Early people learned about iron either

Iron in Nature

Iron occurs naturally in combination with oxygen in the metal ores hematite, limonite, and magnetite. These have about a 50 percent iron content. Since high-grade ores such as these have been mostly exhausted, the use of lower-grade ores such as taconite (with 20 to 40 percent iron content) has increased. In addition, recycled iron and steel have become essential in steel production.

by heating it or while fashioning tools from stone that contained iron. Later, forges were created to smelt or separate the metal from the ore and to eliminate the **carbon** and other impurities before working it into the desired shapes.

Over time, more sophisticated iron production methods were invented. Iron production began in earnest with Abraham Darby's (1678-1717) coke-burning furnaces, which yielded iron in commercial quantities and made it affordable for industry. Darby is considered the father of the Industrial Revolution (England, 1760-1870).

Cast iron is used today for engine blocks, sewer lids, cookware, and fire hydrants.

Pig iron is extracted from iron ore in blast furnaces. It contains about 5 percent impurities, mostly carbon. Most pig iron is sent into steelmaking. It was formerly used to make cast iron, but now cast iron accounts for only about 8 percent of pig iron used.

Cast iron is noted for its strength and its ability to be easily formed, or cast, into shapes. Cast iron was one of the most popular building materials during the 1800s. It is used today for engine blocks, sewer lids, cookware, and fire hydrants.

Wrought iron contains slag, the remains left from the reduction process. It is stronger than cast iron, but is less easily formed. In the past, it has been used for fences, railings, and farm implements, but it is rarely used today.

When iron is alloyed (mixed) with varying amounts of carbon and other metals, a wide range of grades and types of steels are the result. The conversion process invented by Henry Bessemer in 1856 made steel affordable and marked the end of the Iron Age.

See also **Meteor and meteorite; Metabolism**

Irrational number

Rational and irrational numbers make up the system of **real numbers**. Rational numbers can be expressed as the quotient of two integers. That is, rational numbers can be divided into two even halves. Irrational numbers cannot. For instance, the number one can be divided into two halves. That makes one a rational number. Even negatives can be rational numbers if they can be divided into two equal parts.

The ancient Greeks tried very hard to express all numbers as whole numbers. They failed when they came to irrational numbers.

Irrational numbers include π (pi) and the square root of 3, which cannot be expressed as the quotient of two integers. π represents the ratio of the circumference of a circle to its diameter. Mathematically, π is expressed as 3.141592+. The plus at the end means that if divided, the division will go on without end.

The same is true of the square root of 3. To understand this concept, think of the number 9: 3 is the square root of 9 (3 x 3 = 9).

Try to do the same thing with 3. It cannot be done because the square root of 3 is an irrational number.

See also **Decimal system**

Isomer

The concept of isomerism was first noted by French chemist Joseph Gay-Lussac in 1814. He found that the molecular formula for two different compounds can be the same. Because the atoms are connected in different ways, one formula can represent more than one compound. Gay-Lussac concluded that it was the arrangement of a compound's particles that determined the character of a substance.

Isomers are compounds that have identical formulas but different properties due to different atomic structures.

At first, this revolutionary idea was rejected by the scientific community. When Swedish chemist Jöns Berzelius investigated the concept, he found it to be true. Berzelius gave the name "isomers" (from the Greek for "equal parts") to compounds that have identical formulas but different properties due to different atomic structures.

This type of isomer is called a "structural" isomer. In 1848 another type, called an "optical" isomer, or stereoisomer, was found. Since then, scientists have learned that many of the body's important substances, such as **amino acids**, are optical isomers. The existence of nuclear isomers was discovered by Soviet physicist Igor Kurchatov in the 1930s.

Isotope

An isotope is one of two or more atoms that have the same number of protons but different numbers of neutrons.

An isotope is one of two or more atoms that have the same number of protons but different numbers of neutrons.

The concept of isotopes grew out of the study of radioactivity in the late 1800s. Studies of **neon** in the early 1900s revealed that neon appeared to exist in two forms. Scientists thought they had found a new compound of neon or even a new element mixed with the neon. Francis Aston solved this problem with the aid of a new machine, the **mass spectrograph**, that he designed and built. He was able to show that the two forms of neon were actually isotopes of neon. This evidence confirmed the fact that isotopes existed among stable elements and were not peculiar to radioactive elements.

Some Uses of Isotopes in Research

The possible uses of isotopes as research tools became obvious. In 1918 the Hungarian chemist György Hevesy used a radioactive isotope of lead to study the growth of plants. Hevesy added lead to water, which he fed to a plant. He followed the radiation given off by the lead to trace its path through the plant's structures. Hevesy's work was the first in which radioactive isotopes were used as tracers.

In later years, scientists began to use isotopes of elements that occur naturally in plants and animals (lead does not), such as radiocarbon. All living things contain the isotope radiocarbon. When a plant or animal dies, it no longer absorbs radiocarbon from the atmosphere. The radiocarbon already present begins to decay at an exact and uniform rate. Scientists can determine the age of prehistoric objects by measuring their radiocarbon content.

In 1935 the German biochemist Rudolf Schoenheimer (1898-1941) demonstrated that non-radioactive isotopes can also be used as tracers. The use of both radioactive and stable isotopes as tracers is now common in industry, medicine, and research.

See also **Periodic law**

J

Jet engine

To make **aircraft** fly higher and faster, scientists developed heavier **internal combustion engines**. But as the planes flew higher, the thin air at high altitudes reduced engine efficiency and speed. So an even heavier engine that would weigh down the plane was not the answer.

British and Germans Perfect the Jet Engine

Frank Whittle, a British engineer, worked on one possible solution: the jet engine. In 1930 he patented a variation on a gas turbine engine already in commercial use. Meanwhile, the Germans were independently working on a jet engine. They developed the Messerschmitt Me 262, a fighter plane with jet engines that could fly more than 550 miles per hour.

The British were the first to develop the turboprop engine, which added a propeller to the standard jet engine. Their design used an engine developed by Rolls Royce, a company best known for its luxury automobiles. In the late 1950s, Rolls Royce developed a new, more fuel-efficient engine with a large, propeller-like fan at the front.

For the special needs of military aircraft, engineers devised afterburning, which provided a big thrust forward in return for just a small increase in engine weight. In 1950 planes like the United States's F-86 Saber and the Soviet Union's Mig 17 enabled pilots to travel at the speed of sound (Mach 1). Fighters developed in the 1960s and 1970s were capable of speeds of Mach 2 and Mach 3.

Peacetime Uses for Jet Engine Technology

The introduction of jet engines to commercial aviation cut travel time dramatically. After World War II (1939-45), the British produced the world's first jet-powered airliner, the de Havilland Comet, which made its maiden flight in 1949. The Boeing 707, the first widely used jet airliner, revolutionized the industry. A cooperative British and French effort resulted in the *Concorde*, a luxury airliner that used a Rolls-Royce engine to fly at greatly increased speeds.

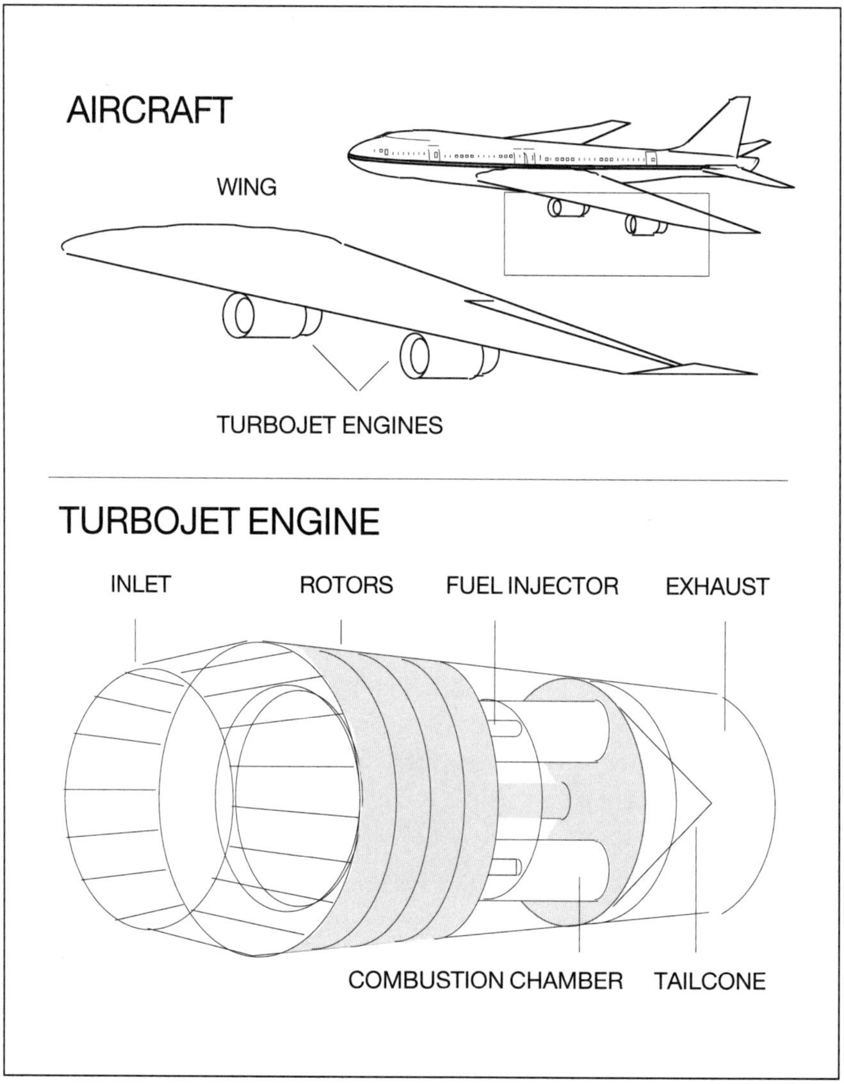

The turbojet engine. The British were the first to develop the turbo engine, which added a propeller to the standard jet engine.

The latest advances in commercial airliners deal with noise reduction, the biggest problem in modern jets. Engineers continue to experiment with noise reduction techniques.

An engine about to be overhauled at the General Electric Aircraft Engine facility in Wichita, Kansas.

Jupiter

Jupiter is the fifth planet from the **Sun** and the largest known planet in the solar system. It is named for the highest god in Roman mythology. At certain periods, it is brighter than any other object in the sky except the Sun and **Venus**. Certain facts concerning the planet have been discovered by a host of astronomers.

Because Jupiter has always been seen by the naked eye, no history of discovery exists for it.

Scientists Study Jupiter

Because Jupiter has always been seen by the naked eye, no history of discovery exists for it.

The first person to detect moons of Jupiter and correctly identify them as satellites of the planet was **Galileo Galilei**, who viewed them through his telescope in 1610. A contemporary of Galileo, Simon Marius (1580-

1624), gave them the mythological names by which they have become known—Io, Europa, Ganymede, and Callisto.

Credit for the discovery of Jupiter's reddish-colored cloud bands has been given to several scientists, including Giovanni Domenico Cassini, an Italian. Cassini determined in the 1600s that the planet's rotation was 9 hours and 56 minutes, a surprisingly short period for such a huge planet.

In 1842 German astronomer Friedrich Bessel calculated the planet's mass. He determined that Jupiter had 388 times the mass of Earth, but had only 1.35 times the density of water, meaning the huge planet was surprisingly light. Bessel thought Jupiter was a white hot planet that continuously heated its swirling clouds.

In the 1870s, Hermann Vogel measured the planet's spectrum and found it was the same as the Sun's. This proved the planet's light was reflected sunlight.

In 1973 Pioneer 10 *explored Jupiter's neighborhood, radioing back an astounding amount of information.*

The idea that **hydrogen** and **helium** were Jupiter's major components was proved correct in 1952 by W. A. Baum and C. A. Code. When Jupiter moved in front of a star, starlight shining through its atmosphere allowed scientists to measure Jupiter with a spectroscope.

A fifth moon, Amalthea, was discovered in 1892 by American astronomer Edward Emerson Barnard. Seth Nicholson is considered the primary discoverer of further moons. Using long-exposure photographs, he detected four additional satellites orbiting Jupiter. By 1951 the total count of the planet's moons was 12.

Jupiter Explored

Five American spacecraft have explored Jupiter's neighborhood, radioing back an astounding amount of information. *Pioneer 10* in 1973 and *Pioneer 11* in 1974 mapped out Jupiter's intense **magnetic field** and gave detailed information about its gravity field. Jupiter was found to be

This dramatic view of Jupiter's Great Red Spot and its surroundings was obtained by Voyager 1 on February 25, 1979. The wavy cloud pattern to the left of the spot is a region of extraordinarily complex and variable wave motion.

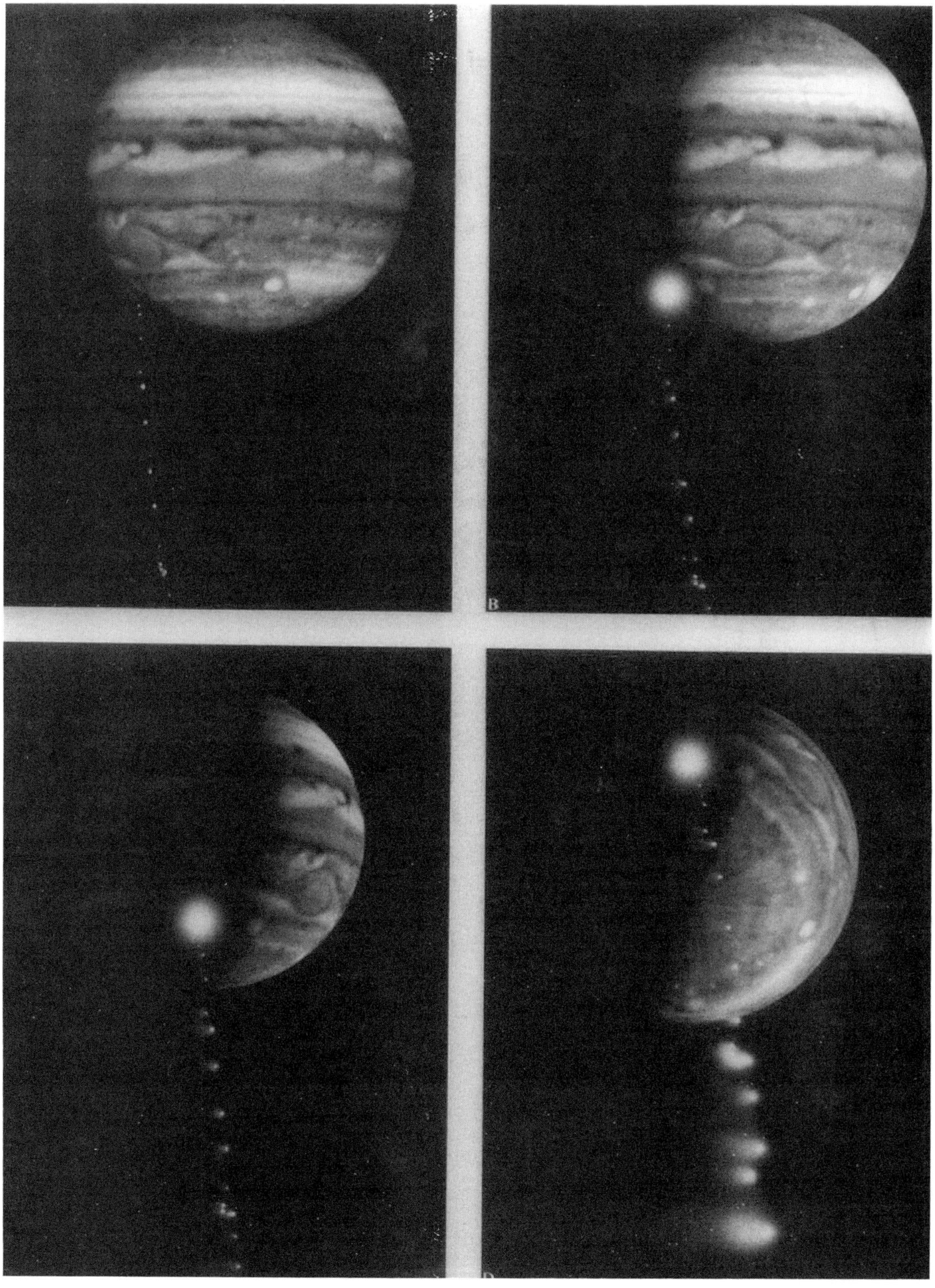

a vast fluid planet with no evidence of a solid surface existing under the perpetual cloud cover.

In 1979 space probes *Voyager 1* and *2* passed by the planet traveling at 45,000 miles per hour (72,405 kph). They sent back spectacular photographs of the Great Red Spot. The probes found that a ring encircles the planet and that volcanic activity and lightning storms occur on Jupiter. They also discovered new moons.

And in July 1994 the Hubble Space Telescope took hundreds of pictures as 20 large chunks of the comet Shoemaker-Levy 9 smashed into Jupiter, raising fireballs more than 1,200 miles wide and scarring the planet with a black dot about half the size of Earth. The images help astronomers learn more about the composition of Jupiter and comets and the dynamics of celestial crashes.

Opposite page: Images of the comet Shoemaker-Levy 9 smashing into Jupiter, taken by the Hubble Space Telescope in July 1994.

Much is still unknown about Jupiter. No one is sure what lies below the clouds and what makes the gaudy colors of these clouds. The spacecraft *Galileo,* on its way to a 1996 rendezvous with Jupiter, will attempt to answer these questions when it circles the planet and drops a probe into the atmosphere.

Master Index

A

Abacists *5:* 904
Abacus *2:* 298; *5:* 904
Abbe, Cleveland *6:* 1067, 1148
Abdomen *4:* 662
Abelson, Philip *4:* 730
Abercromby, Ralph *6:* 1149
Abortion *5:* 907
Abplanalp, Robert H. *1:* 19
Abrasive *1:* **1-2**
Absolute scale *1:* 3
Absolute zero *1:* **2-3**
Acetaminophen *3:* 571
Acetic acid *1:* **3-4,** 7; *5:* 845; *6:* 1170
Acetylcholine *1:* **4-6**; *6:* 1040
Acheson, Edward G. *1:* 1
Acid *1:* 3, 9; *4:* 740
Acid and base *1:* 4, **6-9**; *3:* 563
Acid rain *1:* **9-12**; *4:* 744
Acoustics, physiological *1:* **12-13**
Acquired blindness *2:* 409
Acromegaly *3:* 560
Acropolis *1:* 11
Acrylic plastic *1:* 13-14
Acrylonitrile *3:* 459
ACTH (adrenocorticotropic hormone) *1:* **14-15**; *3:* 553
Actinium *5:* 820
Acupuncture *1:* **15-17**
Adams, John Couch *4:* 728; *5:* 845
Adams' New York Gum *2:* 265
Adams, Thomas, Jr. *2:* 265
Adams, Thomas, Sr. *2:* 265
Addison's disease *3:* 553
Addison, Thomas *3:* 553
Addition *1:* 38; *2:* 227, 274, 281, 294, 299
Additive three-color process *2:* 278
Ader, Clement *6:* 1019
Adhesives and adhesive tape *1:* **17-19**; *2:* 345; *5:* 853, 856
Adiabatic demagnetization *2:* 331
Adipose tissue *3:* 447
Adrenal gland cortex hormones *3:* 552
Adrenalin *1:* 5
Advanced X-ray Astrophysics Facility (AXAF) *6:* 1005
Advertising *3:* 550
Advil *3:* 572
Aëdes aegypti *6:* 1172
Aero Foam *3:* 462
Aerodynamics *6:* 1157
Aeronautics *6:* 1157
Aerosol spray *1:* 19-20
Agate *3:* 444
Agent Orange *2:* 263
Agricultural crops *2:* 336
Agriculture *3:* 453, 499, 500
AIDS (Acquired Immune Deficiency Syndrome) *1:* **20-24**; *2:* 246, 263; *3:* 497, 541, 575; *5:* 848, 893; *6:* 1134
AIDS therapies and vaccines *1:* **24-26**
Aiken, Howard *2:* 300
Air *2:* 241
Airbag, automobile *1:* **26-27**
Air conditioning *1:* **27-29**
Aircraft *1:* 14, 19, **29-36,** 41, 46; *2:* 305; *3:* 459, 585, 594, 535; *4:* 635, 688, 703; *5:* 925, 933, 927; *6:* 1014, 1019, 1035, 1156
Airplane *4:* 761; *6:* 1000
Air pollution *1:* 9; *3:* 514; *4:* 774; *5:* 824
Airship *3:* 564
Airy, George Biddle *2:* 411
alaia *6:* 1036
Alchemy *3:* 562; *6:* 1088
Alcohol *2:* 245; *3:* 450, 509; *6:* 1170, 1171
Alcohol, distilling of *1:* **37-38**
Alcoholic hepatitis *3:* 542
Alcoholism *1:* 37
Aldrin, Edwin "Buzz" *4:* 710; *6:* 984
Aleksandrov, Pavel Sergeevich *6:* 1076
Alexanderson, Ernst *6:* 1058
Algebra *5:* 884
Algorists *5:* 905
Algorithm *1:* **38**; *2:* 227
Alhazen *4:* 653

Boldfaced numbers indicate main entry pages; italicized numbers indicate volume number

Alkaloids *2:* 264
Allen, Horatio *6:* 1083
Allen, Paul *2:* 295
Allergy *1:* **39-40**; *2:* 322
Allotropy *1:* **40**
Alloy *1:* **40-41**
Alpha particle *5:* 878
Alphabet *1:* **41-44**
Alphabet, Chinese *1:* 41
Alphabet, Cyrillic *1:* 41
Alphabet, Roman *1:* 41
Alternating current *1:* **44-46**; *4:* 646
Alternating magnetic field *4:* 752
Alternative energy sources *3:* 451; *6:* 1171
Alternator *1:* 45
Altimeter *1:* **46-48**
Altitude *1:* 46
Aluminum *2:* 290, 384; *4:* 667, 675, 712; *5:* 884, 923, 940, 942
Aluminum foil *3:* 476
Aluminum production *1:* 2, **48-50,** 41
Alvarez, Luis W. *2:* 327
Alvarez, Walter *2:* 327
Alzheimer's disease *4:* 731
Amanita muscaria *3:* 522
Amanita pantherina *3:* 522
Amber *2:* 381
Ambulance *1:* **50-52**
American Arithmetic Company *2:* 229
American Electrical Novelty and Manufacturing Co. *3:* 467
American Sign Language *5:* 923
American Standard Code for Information Interchange (ASCII) *4:* 666; *6:* 1052
Amino acid *1:* 14, **52-53**; *2:* 398; *3:* 498, 593; *4:* 688; *5:* 837, 910; *6:* 1120
Ammonia *4:* 740, 743; *5:* 910; *6:* 1120
Amniocentesis *1:* **53-54**; *5:* 860
Ampère, André *2:* 388
Ampex *6:* 1131
Amplifier *1:* **54-56**; *3:* 528; *4:* 699, 765; *6:* 1019
Amplitude modulation *5:* 921
Amputation *1:* **56-57**; *3:* 479
Anabolism *4:* 688
Anaerobic respiration *3:* 450
Anaglyphic process *6:* 1066
Analog computer *6:* 1041
Analytical engine *2:* 302
Anderson, Carl David *2:* 390; *6:* 1022
Anderson, W. French *3:* 496
Anemia *1:* **57-58**; *2:* 410; *3:* 471; *5:* 811, 892
Anemometers *4:* 635
Aneroid altimeter *1:* 46
Anesthesia *1:* 15, **59-61**; *2:* 277; *3:* 547; *4:* 747, 771; *6:* 1087
Anesthetic *2:* 277
Angier, Robert *3:* 471
Angioplasty, balloon *1:* **61-63**
Angle-closure glaucoma *3:* 508
Angle of incidence *4:* 653
Angle of reflection *4:* 653
Animal breeding *1:* **64-66**; *2:* 276
Antenna *2:* 225; *4:* 701, 702
Anthelin, Ludwig *2:* 358
Anthracycline *2:* 264
Anthrax *3:* 508, 578; *6:* 1064
Antibiotic *1:* **66-71**; *2:* 263, 334; *5:* 848; *6:* 1031, 1044, 1135
Antibodies *1:* 39; *3:* 542, 574; *4:* 731; *5:* 824; *6:* 1043
Antibody and antigen *1:* **71-72**
Antigen *1:* 39; *5:* 892
Antihistamine *1:* 40
Antimacassar *3:* 519
Antimony *6:* 1089
Antineoplastic (anti-cancer) drugs *2:* 264
Antioxidants *3:* 476
Antiparticle *1:* **72-74**; *6:* 1022
Antiproton *6:* 1023
Antiseptic *3:* 478
Anus *2:* 346
Apollo 1 *6:* 983, 984, 1003
Apollo 11 *4:* 711; *6:* 984
Appert, Nicolas François *2:* 234; *3:* 475
Apple Computer Corporation *2:* 309
Appleton, Edward *3:* 590
Appleton layer *3:* 590
Appliances *2:* 276, 348
aqua fortis *4:* 741
aqua regia *1:* 7; *4:* 741
Arber, Werner *5:* 890
Arcades *6:* 1130
Arch *1:* **74-77**
Archer, Frederick Scott *5:* 833
Architecture *2:* 391; *3:* 502
Arc lamp *3:* 467; *6:* 1159
Arculanus *2:* 341
Area digitizers *2:* 348
Argon *2:* 410; *4:* 637, 726; *5:* 883, 914; *6:* 1159
Ariel *6:* 1119
Aristarchus *4:* 709
Aristotle *2:* 342; *3:* 483, 556; *4:* 709
Arithmetic *2:* 227; *5:* 883
Armor *1:* **77-80**; *5:* 909
Armored vehicle *1:* **80-81**; *3:* 467
Armstrong, Neil *6:* 984
Arnold of Villanova *2:* 243
Arrhenius, Svante August *1:* 8; *2:* 243; *3:* 514
Arsenic *5:* 816; *6:* 1089
Arson *3:* 462
Arteriosclerosis *1:* **82-84**; *2:* 410
Arthritis *1:* 14; *2:* 322; *3:* 511, 553
Artificial heart *1:* **84-86**
Artificial intelligence (AI) *1:* **86-89**; *2:* 317, 333, 405; *5:* 900
Artificial ligament *1:* **89-90**
Artificial limb and joint *1:* **90-91**
Artificial skin *1:* **91-93**
Artificial sweetener *1:* **93-94**; *3:* 448
Artillery *1:* **94-96**; *2:* 262; *5:* 902; *6:* 1078
Artwork *2:* 250
Arybhata *2:* 338
Asaro, Frank *2:* 327
Asbestos *2:* 248
Aseptic packaging *3:* 476
Ashton, Francis William *4:* 682
Aspergillus oryzae *3:* 451
Aspirin *3:* 571
Assembly language *4:* 666
Asteroid *2:* 327; *5:* 943
Asteroid belt *4:* 692
Astigmatism *2:* 408, 411
Aston, Francis *3:* 594

Astrolabe *1:* **96-97**; *2:* 293
Astronomical units *6:* 1119
Astronomy *3:* 482; *5:* 841; *6:* 1007, 1114, 1119
Atherosclerosis *2:* 269
Atlantis *6:* 997, 1000
Atmosphere *3:* 588; *4:* 768
Atmospheric circulation *1:* **98-101**; *6:* 1022
Atmospheric composition and structure *1:* **101-104**; *3:* 591; *4:* 694, 774
Atmospheric engine *6:* 1017
Atmospheric pressure *1:* **104-105**; *2:* 339; *4:* 684
Atom *3:* 593; *5:* 810
Atomic and molecular weights *1:* **105-107**; *5:* 821; *6:* 1048
Atomic bomb *1:* 34, **107-111**; *2:* 376; *3:* 567; *4:* 749, 753; *5:* 918
Atomic clock *1:* **111-112**; *2:* 275
Atomic model *6:* 1022
Atomic nucleus *3:* 487; *4:* 752; *6:* 1118
Atomic number *3:* 470; *5:* 821, 858, 912, 918; *6:* 1048, 1068, 1070, 1090, 1118
Atomic theory *1:* **112-114**; *4:* 735; *5:* 867; *6:* 1124
Atomic weight *5:* 845, 860, 912, 918; *6:* 1068, 1070, 1118
Audiocassette *1:* **114-115**; *4:* 669; *6:* 1141
Audiometer *1:* **115-117**
Audiotape *4:* 668
Audio-visual materials *6:* 1046
Auditory brainstem implant *3:* 528
Australopithecus afarensis *3:* 555
Australopithecus ramidus *3:* 560
AutoCAD *2:* 297
Autoclave *6:* 1134
Autofacturing *2:* 307
Autogiro *3:* 535
Autoimmune reaction *3:* 575
Automatic pilot *1:* **117-118**
Automation *2:* 333; *6:* 1138
Automobile *1:* 14, 18, 26; *2:* 252, 317, 329, 398; *3:* 491, 585; *4:* 668; 760; *5:* 926
Automobile, electric *1:* **118-121**
Automobile, gasoline *1:* **121-126**; *3:* 524; *6:* 1028, 1098
Autonomic nervous system *4:* 732
Aviation *1:* 29
Axon *4:* 730, 733
Ayrton, Hertha *2:* 380
AZT *1:* 25; *2:* 263
Aztecs *2:* 266

B

Babbage, Charles *2:* 299, 301, 302
Babbitt, Seward *5:* 897
Baby bottle *1:* **127**
Baby carrier/pouch *1:* **127-128**
Baby food, commercial *1:* **128**
Bacon, Francis *2:* 319; *4:* 653
Bacon, Roger *2:* 410; *3:* 465
Bacteria *1:* **128-132**; *2:* 246, 268; *3:* 472, 499, 504, 508, 575; *4:* 662, 682, 694, 767; *5:* 817, 825, 848, 890; *6:* 1030, 1064, 1108, 1114, 1134, 1171
Bacteriophage *5:* 890
Baekeland, Leo *5:* 842
Baer, Ralph *6:* 1129
Baeyer, Adolf von *5:* 855
Baghouse filters *1:* 11
Baird, John Logie *6:* 1056, 1130
Bakelite *5:* 842
Baker, James *2:* 267
Balloon *1:* 29; *6:* 1147
Balloon frame *5:* 932
Bandage *3:* 478
Bandage and dressing *1:* **132-133**
Band-aid *1:* 18
Banting, Frederick *3:* 552
Barbed wire *1:* **133-134**
Barbiturate *1:* **135**; *3:* 521; *6:* 1086
Bar code *1:* **135-137**; *2:* 256; *4:* 696
Bardeen, John *6:* 1035
Barium *6:* 1163
Barnard, Christiaan *6:* 1092, 1093
Barometer *4:* 684; *6:* 1148
Barr body *5:* 920
Barr, Murray Llewellyn *5:* 920
Bartlett, Neil *6:* 1159
Basal Metabolic Rate (BMR) *4:* 688
Basalt *4:* 712
Basaltic rock *2:* 371; *4:* 706
Base ten *2:* 338
Baseball *1:* **137-138**
BASF *3:* 456
Basketball *1:* **138-140**; *6:* 1086
Bath and shower *1:* **140-143**
Bathyscaphe *6:* 1116
Battery *1:* 45; *2:* 380, 387; *3:* 528, 585; *5:* 940; *6:* 1026
Battery, electric *1:* **143-145**; *3:* 467; *4:* 647
Baudot, Emile *6:* 1052
Baulieu, Etienne-Emile *5:* 907
Bayliss, William *2:* 400; *5:* 914
Bayonet *1:* **145-146**
Bazooka *1:* **146**
BEAM (brain electrical activity mapping) *2:* 384
Beaumont, William *2:* 348
Beauty *3:* 519
Beauty products *2:* 322
Becquerel, Antoine Henri *5:* 877, 881; *6:* 1089, 1117
Bede, Venerable *5:* 922
Bednorz, Georg *6:* 1036
Beebe, Charles William *6:* 1116
Beeckmann, Isaac *3:* 511
Beer *1:* 38
Beeswax *1:* 17
Behaim, Martin *6:* 1046
Behavioral psychology *5:* 814
Behring, Emil von *6:* 1063
Beiersdorf, Paul *1:* 18
Beijerinck, Martinus Willem *6:* 1135
Bell, Alexander Graham *3:* 528, 564; *4:* 636, 698
Bell Telephone Laboratories *2:* 316; *6:* 1019
Bendix *6:* 1142

Boldfaced numbers indicate main entry pages; italicized numbers indicate volume number

Benedictine *1:* 38
Bennett, Floyd *1:* 33
Benz, Karl *3:* 586
Benzene *4:* 656
Beriberi *6:* 1138
Berliner, Emile *4:* 699
Bernard, Claude *2:* 245
Berzelius, Jöns *3:* 593; *5:* 918, 923; *6:* 1121
Bessel, Friedrich *3:* 598
Bessemer, Henry *3:* 593
Best, Charles *3:* 552
Beta particle *5:* 878
Bethe, Hans *3:* 538, 567; *6:* 1032
Beukelszoon, Willem *3:* 474
Bevan, Edward John *5:* 842
B. F. Goodrich Company *6:* 1177
Bich, Marcel *5:* 815
Bichat, Xavier *6:* 1070
Bidwell, Shelford *6:* 1056
Bifocals *2:* 410
Big bang theory *1:* **147-151**; *3:* 512, 527; *5:* 880, 887; *6:* 1013
Bigelow, Erastus *2:* 250
Bigelow, Julian *2:* 333
Bile *2:* 348; *4:* 657; *5:* 916
Binary arithmetic *1:* **151-152**
Binary numbering system *2:* 339; *4:* 665
Binary star *1:* **153-154**; *6:* 1010, 1164
Binomial theorem *1:* **154-155**
Biodegradable plastic *1:* **155-156**
Biological warfare *1:* **156-158**
Biometrics *5:* 855
Biosynthesis *2:* 398
Biot, Jean-Baptiste *4:* 690
Biotechnology *3:* 500
Biotin *6:* 1137
Biplane *1:* 30
Birdseye, Clarence *3:* 475
Birkinshaw, John *6:* 1080
Biro, Georg *5:* 815
Biro, Ladislao *5:* 815
Birth control *1:* **158-161**; *5:* 908
Bismuth *6:* 1089
Bitumen *1:* 17; *5:* 832
Bjerknes, Jacob *6:* 1149
Bjerknes, Vilhelm *4:* 692; *6:* 1149
Blackboard *6:* 1045
Black, G. V. *2:* 341
Black hole *3:* 512, 527; *5:* 841, 868; *6:* 1162, 1164
Black Jack gum *2:* 265
Black, Joseph *2:* 241
Black lights *6:* 1112
Black powder *3:* 465
Bladder shunt *5:* 860
Blaese, R. Michael *3:* 496
Blaiberg, Philip *6:* 1092
Blair, J. B. *2:* 400
Blakeslee, Albert Francis *4:* 722
Bleach *1:* **161-163**
Blériot, Louis *1:* 31
Blibber-Blubber gum *2:* 265
Blind, communication systems for *1:* **163-165**
Blindness *2:* 318; *3:* 508
Bloch, Felix *4:* 752
Bloch, Konrad Emil *1:* 4
Blondel, John F. *2:* 357
Blood *1:* 7, **165-170**; *2:* 245, 394; *3:* 471, 511, 539, 543, 561, 581, 591; *4:* 717, 771; *5:* 812, 891, 933; *6:* 1043, 1094
Blood, artificial *1:* **170**
Blood circulation *3:* 570; *5:* 940
Blood clot dissolving agent *1:* **171**; *2:* 399
Blood group *5:* 892
Blood pressure *2:* 410; *3:* 569; *5:* 939
Blood pressure measuring device *1:* **171-173**
Blood serum therapy *6:* 1064
Blood transfusion *1:* **173-176**
Blood vessels, artificial *1:* **177**
Blow molding *5:* 854
Blue jeans *1:* **178-179**
Blueprints *5:* 827
Blumlein, Alan Dower *6:* 1019
Boat *3:* 564; *6:* 1154
Boeing 707 *3:* 596
Bohr, Niels *5:* 867
Boiler *1:* 37
Bomb *1:* 34; *3:* 564; *5:* 871
Bombardier, Joseph-Armand *5:* 937
Bomber plane *1:* 31
Bondi, Hermann *6:* 1012
Bone allografts *6:* 1091
Boots Company *3:* 571
Borax *5:* 939
Borden, Gail *2:* 236, 318
Borden, John *1:* 18
Borland International Inc. *2:* 295
Borrelia burgdorferi *4:* 661
Bort, Leon Teisserenc de *6:* 1148
Bosch, Carl *3:* 456; *4:* 743
Bottle *2:* 235; *3:* 475; *6:* 1038
Bottling *2:* 234
Boudou, Gaston *3:* 520
Bourbon *1:* 38
Bourne, William *6:* 1024
Boussingault, Jean *4:* 742
Boutan, Louis *6:* 1116
Boveri, Theodor *2:* 271, 404
Bovet, Daniele *1:* 40; *6:* 1031
Boyd, T. A. *3:* 492
Boyer Machine Company *2:* 229
Boyle, Robert *1:* 7; *3:* 565; *4:* 634
Boyle's law *1:* **180**
Boylston, Zabdiel *3:* 576
Bragg, William Henry *6:* 1007, 1166
Bragg, William Lawrence *6:* 1007, 1166
Braham, R. R. *6:* 1149
Brahe, Tycho *4:* 745; *5:* 946
Brahmagupta *5:* 883
Brain *2:* 383
Bramah, Joseph *4:* 659; *6:* 1073
Brand, Hennig *5:* 825
Brandenberger, Jacques E. *5:* 842
Brandy *1:* 38
Branham, Sara Elizabeth *4:* 683
Brass *1:* 41
Braun, Karl Ferdinand *2:* 258; *4:* 646
Braun, Wernher von *2:* 406; *5:* 904
Bread *6:* 1171
Bread and crackers *1:* **180-182**
Breakfast cereal *1:* **182-183**
Breast cancer *4:* 673
Breathed, Berke *2:* 286
Breccia *4:* 712
Breeding *2:* 275
Bréguet, Louis *3:* 537
Brewster, David *4:* 633
Brick *1:* **184-186**
Bridge *1:* **186-189**; *3:* 451

British Adulteration of Food and Drugs Act *3:* 476
British Broadcasting Company (BBC) *6:* 1056
Broadcast band *2:* 280
Broglie, Louis Victor de *5:* 811; *6:* 1099
Bromine *1:* **189-190**; *4:* 772
Bronsted, Johannes Nicolaus *1:* 8
Bronze *1:* 40
Broom, Robert *3:* 559
Brouwer, Luitzen Egbertus Jan *6:* 1075
Brown, Louise *3:* 588
Broxodent *6:* 1074
Brufen *3:* 571
Brunel, Isambard K. *6:* 1098
Brunel, Marc Isambard *6:* 1098
Bruno, Giordano *6:* 1032
Bubble chamber *5:* 873; *6:* 1160
Bubble gum *2:* 265
Bubonic plague *6:* 1064
Budding *6:* 1169
Budding, Edwin *4:* 643
Buell, Kenneth *2:* 345
Buffon, Georges de *2:* 319; *3:* 557
Buildings *3:* 502
Bunsen burner *1:* 190
Bunsen, Robert *6:* 1006
Buoy *1:* **190-191**; *4:* 635
Buoyancy, principle of *1:* **191-193**
Burger, Reinhold *6:* 1123
Burkitt's lymphoma *4:* 663
Burnell, Jocelyn Bell *5:* 863
Burnham, Daniel *3:* 452
Burroughs Corporation *2:* 229
Burroughs, William Seward *2:* 229
Burt, William A. *6:* 1101
Burton, William *3:* 491
Busch, Wilhelm *2:* 284
Buschnel, Noland *6:* 1130
Bush, Vannevar *2:* 293
Bushnell, David *6:* 1024
Butane *3:* 562; *4:* 725
Butter *3:* 447; *6:* 1126
Buttons and other fasteners *1:* **193-194**; *6:* 1177
Byers, Horace *6:* 1149
Byrd, Richard *1:* 33

C

Cables *3:* 459
Cable television *2:* **225-227**; *4:* 704; *6:* 1060
Cacao bean *2:* 266
CAD/CAM *2:* 297, 305, 348
Cadmium *5:* 929
CAE (computer aided engineering) *2:* 307
Cailletet, Louis Paul *4:* 743, 771
Calcium *3:* 470, 570; *4:* 712
Calculable function *2:* **227-228**
Calculating devices *4:* 660
Calculating machines *2:* 298; *6:* 1149
Calculator, pocket *2:* **228-231,** 254; *4:* 644, 645
Calculus *2:* 374; *4:* 738
California *3:* 510
California Institute of Technology (Cal Tech) *2:* 298, 305
Calley, John *6:* 1017
Calne, Donald *3:* 575
Camera *1:* 14; *6:* 1058
Camera obscura *5:* 831
Camras, Marvin *4:* 670
Canal and canal lock *2:* **231-233**; *6:* 1096
Can and canned food *2:* **234-237,** 238
Can opener *2:* **237-238**
Cancer *2:* 245, 263, 265, 267, 322, 336; *3:* 510, 515; *4:* 753; *5:* 812, 878; *6:* 1112, 1162
Canning *3:* 472
Cannon, Walter Bradford *2:* 400; *6:* 1162; 1166
Cantell, Kari *3:* 584
Capacitor *4:* 697
Capp, Al *2:* 286
Carasso, Isaac *6:* 1174
Carbohydrates *2:* 242, 398; *3:* 447, 450, 581; *4:* 656, 688; *5:* 836; *6:* 1169
Carbon 1: 3; *2:* **238-240**; *3:* 561, 563, 592; *4:* 707; *6:* 1033, 1170
Carbon dioxide *2:* **241-242,** 283; *3:* 450, 481; *4:* 725; *5:* 835; *6:* 1128, 1169
Carbon-14 *5:* 878
Carbon monoxide *2:* 239, **243-245,** 330
Carburetor *2:* 398; *3:* 586
Carcinogen *2:* **245-248**
Cargo ship *2:* **248-250**
Carlier, François *3:* 462
Carlson, Chester *5:* 828
Carnallite *5:* 858
Carnival *3:* 451
Carothers, Wallace Hume *2:* 335; *4:* 755; *5:* 843, 851, 852
Carpet *2:* **250-252**; *4:* 643, 756; *5:* 854, 855; *6:* 1144
Carrel, Alexis *6:* 1091
Carrier *3:* 541
Carrier, Willis H. *1:* 28
Carson, Rachel *2:* 338
Car wash, automatic *2:* **252**
Caselli, Abbe *6:* 1056
Cash register *2:* **253-254**; *2:* 309; *4:* 696, 699
Cassettes *4:* 668
Cassini, Giovanni Domenico *3:* 598; *6:* 1128
Castner, Hamilton Young *5:* 941
CAT (computerized axial tomography) *5:* 808, 809
Catabolism *4:* 688
Catalysts *2:* 398; *5:* 845, 890
Catalysts and catalysis *2:* 399
Catalytic converters *3:* 492
Cataract surgery *2:* **254-255**
Cataracts *2:* 254, 319, 409; *3:* 508, 515
Catastrophe theory *6:* 1076
Catheter, cardiac *2:* **256**
Cathode ray *6:* 1160
Cathode-ray tube *2:* **256-258,** 293, 308; *6:* 1165
CATV (Community Antenna Television) *2:* 225
Cavendish, Henry *3:* 512, 565
Cavities *3:* 468
Cavitron *6:* 1110
Cayley, George *4:* 636; *6:* 1156
CD-ROM *6:* 1047
C-curity *6:* 1176

Boldfaced numbers indicate main entry pages; italicized numbers indicate volume number

Cell 1: *4, 24;* 2: ***258-260,*** *269, 276, 352, 398;* 3: *471, 493, 539, 570, 574;* 4: *656, 717, 720, 730, 733, 741, 752;* 5: *824, 890, 912, 914;* 6: *1039, 1044, 1069, 1134, 1169*

Cell division 2: *270;* 3: *546;* 5: *920*

Cellophane *5:* 842; *6:* 1014

Cell staining *2:* 270; *4:* 733

Cell theory *6:* 1070

Cellular telephones *5:* 876, 921

Celluloid *5:* 834, 842; *6:* 1152, 1169

Central heating *3:* 533

Central nervous system *4:* 731

Central processing unit *4:* 699

Ceramic *3:* 445

Ceremonies *3:* 465

Cervical cancer *5:* 807

Cesium *6:* 1007

Chadwick, James *5:* 862; *6:* 1022

Chaffee, Roger *6:* 984

Chain, Ernst *5:* 817

Chalkboard *6:* 1045

Challenger *5:* 904; *6:* 985

Chamberlain, Owen *6:* 1022

Chamberland, Charles-Edouard *6:* 1134

Chandrasekhar, Subrahmanyan *4:* 745

Chang Heng *5:* 918

Chappe, Claude *6:* 1083

Chardack, William *5:* 805

Chardonnet, Comte Louis de *5:* 842

Charles, Jacques-Alexandre-César *1:* 2; *3:* 565

Chemical bonding *6:* 1125

Chemical element *1:* 40; *2:* 390; *3:* 565; *5:* 845; *6:* 1070, 1088, 1124

Chemicals *2:* 336; *3:* 467, 470, 509

Chemical warfare *2:* 260-263

Chemistry *1:* 6; *2:* 331, 334, 335; *3:* 447, 453, 459, 564, 593; *4:* 687, 765; *5:* 818, 826, 853, 881, 906; *6:* 1088, 1124, 1143, 1159

Chemotherapy *2:* **263-264**; *6:* 1044

Chernobyl *5:* 878

Chewing gum *2:* **264-266**

Chicken pox *3:* 542; *5:* 848; *6:* 1134

Chicle *2:* 264

Chiclets *2:* 265

Children of a Lesser God *5:* 923

Chimera *5:* 890

China *3:* 465; *4:* 635

Chinese medicine *1:* 15

Chiron Corporation *3:* 497

Chlorine *2:* 284; *3:* 563; *4:* 685, 772; *5:* 941

Chlorofluorocarbons (CFCs) *1:* 20; *3:* 470; *4:* 773; *5:* 855

Chloroform *4:* 656

Chlorophyll *2:* 242; *4:* 668; *5:* 835; *6:* 1169

Chlorpromazine *6:* 1087

Chocolate *2:* **266-268**

Chocolate chip cookie *2:* 267

Choh Hao Li *1:* 14

Cholecystokinin (CCK) *5:* 916

Cholera *2:* **268-269**; *3:* 508, 574, 578

Cholesterol *2:* **269**; *1:* 4; *3:* 447, 570; *4:* 657

Christy, James *5:* 847

Chromatin *2:* 259

Chromium *4:* 674

Chromosome *2:* 259, **269-271,** 352, 405; *3:* 493, 543; *4:* 720; *5:* 919

Chromosphere *6:* 1033

Chronometer *4:* 725

Church, Alonso *2:* 228

Church-Turing thesis *2:* 228

Chyme *2:* 347

Cigarettes *2:* 245

Cinématographe *4:* 715, 716

Cinnabar *4:* 686

Cipher *6:* 1175

Circulation *2:* 256

Circulatory system *3:* 529

Cirrhosis *3:* 541

Citizen's band *5:* 876

Clad fiber *3:* 457

Clark, Alvan *6:* 1055

Clarke, Arthur C. *2:* 286

Clasp locker *6:* 1176

Classical physics *5:* 867, 888; *6:* 1099

Claude, Georges *4:* 727

Clepsydra *2:* 272

Climate changes *3:* 515

Climatology *6:* 1147

Clock *4:* 644; *5:* 897; *6:* 1066

Clock and watch *2:* **271-275,** 340; *3:* 467; *4:* 658; *6:* 1078

Cloning *2:* **275-276**

Closed loop *5:* 886

Clostridium tetani *6:* 1063

Cloth *2:* 335

Clothes dryer *2:* **277**; *4:* 642, 699

Clothing *2:* 357; *3:* 459; *6:* 1141

Cloud chamber *5:* 873

Coal *2:* 243; *3:* 561; *4:* 694; *5:* 822

Coal mining *2:* 248

Coaxial cable *2:* 227

Coca plant *2:* 277

Cocaine *2:* **277-278**; *4:* 747; *6:* 1062

Cochlea *1:* 13

Cochlear implants *3:* 528

Cochrane, Josephine *2:* 349

Codon *3:* 498

Coenzyme *6:* 1137

Cohen, Stanley *4:* 731

Colchicine *2:* 264

Cold permanent wave *3:* 520

Cold war *2:* 377

Collins, Michael *6:* 984

Collodion *5:* 833, 842

Color blindness *2:* 409

Color photography *2:* **278-280**; *5:* 834

Color spectrum *2:* **280-282**; *4:* 656; *5:* 886, 887

Colossus *2:* 300

Colt, Samuel *4:* 679

Columbia *2:* 277

Columbia *6:* 985, 999

Columbia Broadcasting System (CBS) *6:* 1058

Columbian Exposition *1:* 46

Columbus, Christopher *2:* 266, 291; *3:* 510; *6:* 1042, 1046

Combinatorial topology *6:* 1075

Combustion *2:* 241, **282-284**; *4:* 768

Comet *2:* 240; *3:* 512; *4:* 691; *5:* 840, 943, 947; *6:* 1119

Comet *1:* 36; *3:* 596

Comic strip and comic book *2:* **284-286**; *5:* 927
Communication *2:* 310, 315, 348; *3:* 449, 456; *4:* 639
Communications satellite *2:* 227, **286-289**; *4:* 704; *6:* 1055
Compact disc player *2:* **289-290**; *4:* 699
Compact discs *6:* 1019, 1141
Compass *2:* **290-292,** 371, 386; *4:* 725
Compound *2:* 238, 241; *3:* 593
Compressed air illness *2:* 339
Compressed yeast *6:* 1171
Compression and transfer molding *5:* 854
Compression wave *6:* 1107
Comptometer *2:* 229
Compton, Arthur Holly *4:* 655
Computer *1:* 38; *2:* 293, 331, 348, 384; *3:* 551; *4:* 644, 695; *5:* 900
Computer, analog *2:* **293-294,** 302; *4:* 697
Computer application *2:* **294-297,** 307
Computer art *2:* 317
Computer-Assisted Instruction (CAI) *6:* 1047
Computer chip *5:* 869
Computer, digital *2:* 227, 254, 292, **298-302,** 305, 308, 312, 316, 333, 406; *3:* 450; *4:* 665, 697; *5:* 880, 900; *6:* 1079, 1130
Computer disk and tape *2:* **303-305,** 310; *4:* 697; *5:* 836
Computer graphics *2:* 348
Computer, industrial uses of *2:* **305-307,** 312, 316, 333, 348; *5:* 900
Computer input and output device *2:* 307, **308-310,** 311, 312, 316, 348; *4:* 666, 697; *5:* 933; *6:* 1130, 1139
Computer Integrated Manufacturing *2:* 307
Computer network *2:* **310-311,** 312; *4:* 697
Computer operating system *2:* 305, **311-312,** 333; *4:* 666; *6:* 1035
Computer output microfilm *2:* 310
Computer scientist *2:* 302
Computer simulation *2:* 307, **313-315**
Computer speech recognition *2:* 310, **315-316,** 348; *6:* 1130
Computer storage *3:* 551
Computer technology *2:* 405
Computer vision *2:* 307, 310, **316-317**
Concave lenses *4:* 648
Concentrated fruit juice *2:* **317-318**
Concord Academy *6:* 1045
Concorde *1:* 36; *3:* 596
Concrete and cement *2:* 368; *6:* 1084
Condensed milk *2:* 236
Condenser *1:* 37
Conditioned reflex *5:* 814
Conduction *3:* 532
Congenital blindness *2:* 409
Conjunctivitis *2:* 409
Contact lens *2:* **318-319,** 412
Containers *2:* 401
Contaminant *2:* 336
Continental drift *2:* **319-322,** 370; *4:* 758; *5:* 843, 844, 880; *6:* 1117
Contraception *5:* 908
Convection *3:* 532
Convex lenses *4:* 648
Conveyor belt *2:* 349; *5:* 829
Cookie *2:* 266
Coolidge, William *6:* 1166
Cooper, Leon *6:* 1035
Cooper, Peter *6:* 1081
Copernicus, Nicolaus *3:* 482, 484; *5:* 838, 943; *6:* 1032
Copper *2:* 384; *3:* 538, 591; *4:* 758; *6:* 1068, 1088
Coral Draw *2:* 297
Corliss, George *6:* 1018
Cornea *2:* 318
Corneal contact lenses *2:* 318
Corneal sculpting *5:* 872
Corning Glass Works *3:* 456
Corning, Leonard *2:* 278
Corona *6:* 1033
Coronagraph *5:* 947
Coronagraphic photometer *5:* 947
Corpuscular theory *4:* 654
Corrosive *1:* 3
Cortisone *2:* **322,** 269; *3:* 552; *4:* 657
Cosmetics *2:* **322-326**; *3:* 519, 520; *4:* 764
Cosmic ray *3:* 488; *5:* 866, 948; *6:* 1022, 1111, 1125, 1163
Cosmological principle *6:* 1012
Cosmos *5:* 912
Cosmos and Damian *6:* 1090
Cowcatcher *6:* 1081
Cowen, Joshua *3:* 467
Cowpox *3:* 576
Crab nebula *3:* 488
Crack *2:* 278
Cramer, Stuart W. *1:* 28
Crane *2:* 249; *5:* 897
Crapper, Thomas *6:* 1073
Cray, Seymour *6:* 1034
Creed, Frederick G. *6:* 1051
Crest *6:* 1075
Cretaceous catastrophe *2:* **326-329**
Crick, Francis *2:* 354, 404; *3:* 547
Crime *2:* 277; *3:* 460, 501
Crime prevention *4:* 649, 657
Criminal *4:* 649
Cro-Magnon man *3:* 558
Crookes, William *4:* 743; *6:* 1064
Crops *2:* 336
Cross, Charles Frederick *5:* 842
Cruise control, automobile *2:* **329-330**
Cryogenics *2:* **330-331**; *3:* 538
Cryolite *3:* 470
Cryopreservation *3:* 588
Cryosurgery *2:* 255
Crystal *5:* 877; *6:* 1162
Crystallography *2:* 355; *6:* 1162
Crystal rectifiers *4:* 646
Cummings, Alexander *6:* 1073
Cure *2:* 331
Curie, Marie *2:* 331; *5:* 877, 881
Curie, Pierre *2:* 331; *5:* 837, 877, 881

Boldfaced numbers indicate main entry pages; italicized numbers indicate volume number

Curium *2:* **331-332**
Curling iron *3:* 519
Currents *4:* 758
Cursive *1:* 44
Cushing, Harvey *2:* 278; *3:* 553
Cushing's disease *3:* 553
Cutting tools *4:* 640
Cybernetics *2:* 227, 307, **332-333**; *4:* 666; *5:* 897
Cyclones *6:* 1021
Cyclosporin *6:* 1093
Cyclotron *6:* 1167
Cyril, Saint *1:* 44
Cystic fibrosis *2:* **333-334**; *3:* 496, 500
Cytoplasm *2:* 258, 260

D

Dacron *2:* **335-336**; *3:* 460; *5:* 852
Daguerre, Louis-Jacques-Mande *5:* 832
Daguerreotype *5:* 832
Daimler, Gottlieb *3:* 586
Dale, Henry *1:* 6
d'Alembert, Jean Le Rond *6:* 1145
Dalrymple, Brent *2:* 370
Dalton, John *6:* 1022, 1124
Damadian, Raymond *4:* 752
Damage *2:* 364
Danzer, John *5:* 827
Darby, Abraham *3:* 592
d'Arcoli, Giovanni *2:* 341
Dark matter *3:* 512
Darting gun *3:* 525
Dart, Raymond A. *3:* 559
Darwin, Charles *2:* 402; *3:* 544, 554; *4:* 721; *5:* 812
Data *2:* 308, 310; *4:* 671
Database *2:* 302
Davidson, Barry *6:* 1143
Davies, J. A. V. *6:* 1043
Davis, K. H. *2:* 316
Davis, Phineas *6:* 1081
Davy, Humphry *2:* 380; *5:* 857, 940
Dawson, Charles *3:* 558
Day, Henry *1:* 18
DDT (dichloro-diphenyl-trichloroethane) *2:* **336-338**
Deafness *5:* 921
Debye, Peter *2:* 331
Decimal system *2:* 300, **338-339**; *3:* 593
Decompression chamber *2:* 340
Decompression sickness *2:* **339-340**
Dedekind, Richard *5:* 884
Deficiency diseases *6:* 1138
De Forest, Lee *4:* 715
d'Elhuyar, Don Fausto *6:* 1095
Deluc, Jean Andre *6:* 1148
Dendrite *4:* 733
Dental drill *2:* **340-341**
Dental filling, crown, and bridge *2:* **341-342**
Dentistry *2:* 340, 341; *3:* 470
Derrick *4:* 762
Desalination techniques *2:* **342-343**
Descartes, René *4:* 647
The Descent of Man *3:* 557
Deslandres, Henri-Alexandre *5:* 946
Determinism *5:* 889
Deuterium *3:* 565
Devol, George *5:* 898
Dewar flask *2:* 330
Dewar, James *2:* 330; *3:* 565; *6:* 1123
Dextran sulfate *1:* 25
Diabetes *2:* 334, 410; *3:* 499, 552, 583
Diagnosis *2:* 382, 383, 393, 405; *3:* 471; *4:* 670, 673, 751; *5:* 806
Dialogue Concerning the Two Chief World Systems—Ptolemaic and Copernican *3:* 485
Dialysis machine *2:* **343-344**
Diamonds *1:* 40; *2:* 238
Diaper, disposable *2:* **344-345**
Diastase *2:* 399
Dickinson, J. T. *2:* 335
Dickson, William Kennedy Laurie *4:* 714, 716
Diesel engine *6:* 1028, 1084
Diesel fuel *4:* 761
Diesel, Rudolf *2:* 397; *6:* 1026
Difference engine *2:* 299
Differential *2:* 293
Diffraction grating *2:* 282; *6:* 1007
Diffuse reflection *4:* 652
Digestion *2:* **345-348,** 394, 398, 399; *3:* 562, 581; *5:* 812, 876, 914, 915; *6:* 1138, 1170
Digestive tract *2:* 346
Digital audio tape (DAT) *4:* 669, 745
Digital computer *2:* 348; *4:* 671, 765
Digital instruments *6:* 1041
Digital transmission *3:* 458
Digitizer *2:* 310, **348**
Digitizing tablet *2:* 308, 348
Dinoflagellate *5:* 887
Dinosaurs *2:* 327
Diode *2:* 290
Dioxins *5:* 851
Diphtheria *3:* 578; *6:* 1064
Dirac, Paul Adrien Maurice *2:* 389; *5:* 866; *6:* 1022
Direct current *1:* 44; *4:* 646
Dirigible *1:* 29
Dirks, Rudolph *2:* 284
Discontinuity *2:* 371
Discovery *6:* 1000
Disease *1:* 15; *2:* 245, 263, 268, 322, 331, 333; *3:* 447, 504, 508, 510, 539, 541, 573, 575, 581, 583; *5:* 849; *6:* 1063
Dishwasher *2:* **349-350**
Disinfectants *6:* 1063
Distillation *2:* 342
Diving *2:* 339
Diving apparatus *2:* **350-352**
Division *1:* 38; *2:* 227, 228
DNA (deoxyribonucleic acid) *2:* 259, 269, **352-356,** 404, 405; *3:* 498, 499, 547; *5:* 893, 910; *6:* 1162, 1166
Döbereiner, Johann *5:* 819
Dobzhansky, Theodosius *2:* 404; *4:* 722
Documents *3:* 449
Doell, Richard *2:* 370
Dog biscuit *2:* **356**
Dolby, Ray *4:* 744
Dolby system *4:* 744
Dollond, John *6:* 1054
Domagk, Gerhard *6:* 1030
Dominant gene *3:* 494
Donald, Ian *6:* 1109
Donkin, Bryan *2:* 235
Donovan, Marion *2:* 344
Doppler, Christian Johann *5:* 886

Doppler effect *5:* 887
Doppler radar *5:* 872; *6:* 1149
Dorn, Friedrich *5:* 883
Dot matrix *2:* 310
Doughnut *2:* **356-357**
Douglas, J. Sholto *3:* 569
Downs cell *5:* 941
Downsizing *4:* 673
Down's syndrome *5:* 860
Drake, Edwin L. *4:* 762
Drake, Jim *6:* 1154
Draper, John *5:* 826
Drebbel, Cornelius *6:* 1024
Dresler, Heinrich *3:* 547
Drew, Richard *1:* 18
Drill *4:* 762; *6:* 1097
Drillstring *4:* 763
Drinker, Philip *5:* 849
Drinker tank respirator *5:* 849
Dripps, Isaac *6:* 1080
Driving *2:* 329
Droperidol *6:* 1088
Dr. Scott's Electric Toothbrush *6:* 1074
Dr. Sheffield's Creme Dentifrice *6:* 1075
Drug trade *2:* 277
Drugs *2:* 277, 393; *3:* 521, 547, 571
Dry cleaning *2:* **357-359**
Dry yeast *6:* 1171
Du Fay, Charles *2:* 381
Du Pont *1:* 28; *4:* 755
Dubble Bubble gum *2:* 265
Dubois, Marie-Eugene *3:* 558
Duchenne, Guillaume B. A. *4:* 718
Duchenne muscular dystrophy *3:* 500; *4:* 718
Duke, Sam *6:* 1151
Dunkin' Donuts *2:* 357
Duodenum *5:* 914
Durand, Peter *6:* 1068
Dwarfism *3:* 560
Dynamite *2:* **359-361**; *5:* 909; *6:* 1097
Dystrophin *4:* 718

E

Eagle *6:* 984
Ear canal *1:* 12
Ear drum *1:* 12
Earhart, Amelia *1:* 33
Ear mold *3:* 528
Earth *2:* 242; *3:* 481, 588; *4:* 677, 723, 737, 771; *6:* 1114
Earthquake *2:* **363-364,** 371; *4:* 704, 706; *5:* 843, 844, 883, 916
Earthquake measurement scale *2:* 364, **365-366,** 368
Earthquake-proofing techniques *2:* 364, **366-368**
Earth's core *2:* **368-370,** 371, 372; *4:* 707; *5:* 843, 918; *6:* 1116
Earth's magnetic field *2:* **370-371,** 386; *4:* 758; *5:* 949
Earth's mantle *2:* 368, 370, **371-372**; *4:* 706, 758; *5:* 918
Earth survey satellite *2:* **372-374**; *4:* 704
Earth tremors *2:* 364
Ear trumpet *3:* 528
Eastman Dry Plate Company *5:* 834
Eastman, George *2:* 279; *4:* 680, 714, 716; *5:* 834
Easy Washing Machine Company *6:* 1142
EBCDIC *4:* 666
Ebola fever *1:* 22
Eccles, John Carew *1:* 6
Echo *5:* 880
Echolocation *5:* 869
Echo sounder *4:* 757
Eckert, J. Presper *2:* 303
Eclogite *2:* 372
Ecology *3:* 481
Ecosystem *2:* 336; *3:* 471
Edison, Thomas Alva *1:* 45; *4:* 699, 714, 716; *6:* 1162, 1166
Ehrlich, Paul *2:* 263; *6:* 1043, 1044
Eight-track cartridges *4:* 669
Eilenberg, Samuel *6:* 1076
Einhorn, Alfred *4:* 747
Einstein, Albert *2:* **374-377**; *3:* 512, 527; *5:* 829, 841, 867, 888, 889
Einsteinium *2:* **377**
Einthoven, Willem *2:* 382
Eisenhart, Luther *6:* 1076
Eisenhower, Dwight D. *5:* 895; *6:* 981
Elastic limit *2:* 379
Elasticity *2:* **378-379**
Elastomers *5:* 853
Electrical induction *1:* 45
Electric arc *2:* **380-381**; *4:* 637
Electric blanket *2:* **381**; *3:* 532
Electric charge *2:* **381-382**
Electric curling iron *3:* 520
Electric fan *3:* 446
Electric hot-air dryer *3:* 520
Electricity *1:* 8, 45; *2:* 257, 381, 385, 387; *4:* 635, 644, 645, 694, 702, 703, 728, 771
Electric light *2:* 380; *3:* 523
Electric motor *2:* 393; *6:* 1156
Electric organ *4:* 720
Electrocardiogram *2:* 382
Electrocardiograph (ECG) *2:* **382-383**
Electrodynamics *2:* 388
Electroencephalogram (EEG) *2:* **383-384**
Electrogasdynamics *4:* 725
Electrolysis *2:* **384-385**; *5:* 857, 919, 940
Electromagnet *4:* 702; *6:* 1050
Electromagnetic field *5:* 866
Electromagnetic spectrum *5:* 880
Electromagnetic wave *2:* **385**; *3:* 487; *4:* 655, 700; *5:* 874; *6:* 1111, 1161, 1165
Electromagnetism *2:* 376, 377, 382, **386-388**; *3:* 459, 548; *4:* 651, 656; *5:* 827, 830, 867, 929; *6:* 1068, 1095, 1108, 1114, 1145
Electron *2:* 382, **388-390**; *3:* 564; *4:* 656, 709, 726; *5:* 829, 841, 863, 866, 867; *6:* 1022, 1090
Electronic *6:* 1141
Electronic mail *6:* 1051
Electronic mailbox *6:* 1049
Electronics *2:* 277, 294; *4:* 645, 745; *6:* 1129, 1131

Boldfaced numbers indicate main entry pages; italicized numbers indicate volume number

Electron microscope *2:* 256; *3:* 551
Electrostatic precipitators *1:* 11
Electrostatics *2:* 388
Element *2:* 238; *3:* 470, 487, 509, 530, 537, 564; *4:* 730, 767; *5:* 819, 918, 939; *6:* 1032, 1048, 1064, 1068, 1070, 1088, 1095, 1117, 1118, 1124, 1159
Element 95 *2:* 331
Element 96 *2:* 331
Elements 104-109 *2:* **390-391**; *5:* 821
Elevator *2:* **391-393**; *4:* 696; *5:* 932
Eliason, Carl *5:* 937
Elion, Gertrude Bell *2:* 263
Elizabeth I *3:* 441; *6:* 1072
Elliott, Thomas R. *1:* 5; *6:* 1040
Elsener, Karl *6:* 1038
E-mail *2:* 297, 310
Emulsified starch *3:* 448
Endeavour *6:* 985, 1000
Enders, John *4:* 717
Endocrinology *6:* 1094
Endorphin and enkephalin *2:* **393**; *3:* 553
Endorphins *1:* 17
Endoscope *2:* **393-394**; *3:* 457
Energy *2:* 345; *3:* 467, 487, 491, 530
Engine *1:* 30; *2:* 349, **394-396**; *3:* 537; *5:* 937
Engine oil *2:* **396-398**
Engineering *2:* 294, 305, 359
English *4:* 639
ENIAC *2:* 300
Enterprise *6:* 999
Entertainment *2:* 286, 312; *3:* 465
Entropy *3:* 531
Environment *1:* 9
Environmental Protection Agency (EPA) *1:* 9, 11; *4:* 643; *5:* 883
Enzyme *2:* 346, **398-399**; *3:* 450, 591; *5:* 890, 919; *6:* 1137, 1170
Epicenter *2:* 364, 365
Epidemic parotitis *4:* 717
Epilepsy *6:* 1008
Epinephrine (adrenaline) *2:* **399-400**; *3:* 552
Epstein-Barr *3:* 542
Eraser *2:* **400**; *5:* 906; *6:* 1104
Erlich, Henry *3:* 502
Ernst, Richard R. *4:* 753
Erythroblastosis fetalis *5:* 892
Esaki, Leo *6:* 1099
Esophagus *2:* 347
Estridge, Philip D. *2:* 304
Ethanol *3:* 451; *6:* 1171
Ether *4:* 656
Ethics *3:* 500
Ethylene *2:* **400-401**
Euler, Leonhard *6:* 1075, 1145
Euphonia *6:* 1138
Evans, Daniel *3:* 444
Evans, Oliver *4:* 679; *6:* 1018
Everest, Herbert *6:* 1151
Evidence *4:* 650
Evolution *3:* 493; *4:* 720
Evolutionary theory *2:* **401-405**; *3:* 560; *6:* 1117
Ewing, William Maurice *4:* 704; *6:* 1116
Expert system *2:* **405-406**
Explorer 1 *2:* **406-407**; *5:* 904
Explorer VI *6:* 1150
Explosives *2:* 359; *5:* 909
Extinction *2:* 327
Extrusion *5:* 854
Eye *2:* 254
Eye disorders *2:* 318, **407-410**; *3:* 508; *5:* 872
Eyeglasses *2:* 318, **410-412**; *4:* 647; *6:* 1055

F

Faber, Joseph *6:* 1138
Fabry, Charles *4:* 773
Factor VIII *3:* 541
Faget, Maxim *6:* 998
Fahrenheit, Gabriel Daniel *3:* 530; *4:* 684
Fair *3:* 451
Fairfax, Virginia *5:* 895
False teeth *3:* **443-445**
Familial hypercholesterolemia *3:* 497
Fan *1:* 27; *3:* **445-446**; *6:* 1156
Faraday, Michael *1:* 8, 45; *2:* 257, 387, 388
Farman, Henri *1:* 31
Farming *3:* 453
Farnese globe *6:* 1046
Farnsworth, Philo T. *2:* 258; *6:* 1058
Farsightedness *2:* 407, 410
Fashion *3:* 445
Fastener *6:* 1176
Fat *2:* 398; *3:* 560, 583; *4:* 688
Fat Man II *3:* 466
Fats *4:* 656, 717
Fats and oils *3:* **447**
Fat substitute *3:* **448-449**
Fauchard, Pierre *2:* 340, 341
Fault lines *2:* 363
Fax machine *3:* **449-450**
Federal Communications Commission (FCC) *6:* 1058
F-86 Saber *3:* 595
Feit, Louis *3:* 521
Felt, Eugene *2:* 229
Felt-tip pen *5:* 815
Female disease *4:* 673
Fender, Clarence *4:* 720
Fermentation *2:* 241, 399; *3:* **450-451,** 474, 505; *5:* 817; *6:* 1169
Fermented *1:* 37
Fermi, Enrico *4:* 748, 753; *5:* 866
Ferrin, Otis *6:* 1143
Ferris, George Washington Gale *3:* 451
Ferris wheel *3:* **451-453**
Fersman, Aleksandr *4:* 773
Fertilization *3:* **453-454**
Fertilizer *4:* 764; *5:* 825, 858
Fertilizer, synthetic *3:* **454-456**
Festivals *3:* 465
Fey, Charles *5:* 936
Fiber *4:* 633; *5:* 855
Fiberglass *3:* **456,** 478; *6:* 990, 1037, 1154
Fiber optics *2:* 227, 289, 311, 394; *3:* **456-459**; *4:* 652; *6:* 1051
Fiber, synthetic *1:* 13, 14; *2:* 252, 336; *3:* **459-460**
Fiberscope *2:* 393; *3:* 457
Fichet, Alexander *5:* 909
Fick, Adolf *2:* 318
Fighter plane *1:* 34
Figuring *2:* 298
Film *3:* 580
Film badge *5:* 873

Finance *3:* 549
Fingerprinting *3:* **460-462**
Finlay, Carlos Juan *6:* 1172
Fire extinguisher *3:* **462,** 464
Firefighting *3:* 488
Firefighting equipment *3:* **462-464**
Fireproofing techniques *2:* 367
Fireworks *3:* **465-466**; *5:* 857, 859
First law of thermodynamics *3:* 531
Fischer, Hans *2:* 279
Fisher, Bud *2:* 284
Fisher, E. G. *2:* 293
Fisher, Ronald *2:* 404
Fission (atomic) bombs *4:* 749
Fistula *2:* 344
Fixed air *2:* 241
Flamethrower *3:* **466-467**
Flammarion, Camille *5:* 846
Flamsteed, John *6:* 1118
Flashbulbs *6:* 1160
Flashlight *3:* **467**
Fleer, Frank *2:* 265
Fleming, Alexander *5:* 816
Fleming, John Ambrose *4:* 646
Flemming, Walther *2:* 270; *3:* 546
Flight *1:* 29; *3:* 535, 596; *4:* 636
Flight training *2:* 312
Floppy disk *2:* 303
Florey, Howard *5:* 817
Fluorescence and phosphorescence *3:* **467-468**; *4:* 728; *5:* 826; *6:* 1044, 1114, 1167
Fluorescent *6:* 1043
Fluoride *3:* 470; *6:* 1075
Fluoride treatment, dental *3:* **468-469**
Fluorine *2:* 284; *3:* **470-471**; *6:* 1159
Fluorocarbon *1:* 28
Fluorspar *3:* 470
Flying *1:* 29
Flywheel *6:* 1018
Foaming *5:* 854
Foam rubber *6:* 1037
Focke, Heinrich *1:* 33
Fokker, Anthony *1:* 31
Folic acid *3:* **471**
Folling, Asbjorn *5:* 837
Food *2:* 234, 266, 346, 356; *3:* 447, 450, 524
Food additive *3:* 509
Food chain *1:* 10; *2:* 336; *3:* **471-472**; *4:* 741
Food preparation *2:* 398
Food preservation *2:* 234, 331; *3:* **472-476**
Food spoilage *3:* 473
Foot-and-mouth disease *6:* 1135
Football *3:* **476-478**
Ford, Henry *3:* 586; *4:* 680
Ford Motor Company *2:* 333
Forensic science *3:* 501
Forlanini, Enrico *3:* 564
Forssmann, Werner *2:* 256
Fossil fuel *3:* 514; *4:* 723
Fossils *3:* 555
Fourier, Jean-Baptiste-Joseph *6:* 1145
Foyn, Svend *3:* 525
Fracastoro, Girolamo *6:* 1042
Fractions *5:* 883
Fractures, treatments and devices for treating *3:* **478-479**
Frankland, Edward *6:* 1124
Franklin, Benjamin *2:* 381, 387, 411; *3:* 534; *4:* 635
Fraunhofer diffraction *4:* 650
Fraunhofer, Joseph von *6:* 1006
Fréchet, Maurice-René *6:* 1076
Frederick the Great *2:* 341
Freedom *6:* 1003
Freedom 7 *6:* 981
Freeze-dried *3:* 581
Freezing *2:* 331; *3:* 472
French Sign Language *5:* 923
Freon *1:* 20, 28
Frequency *2:* 385; *5:* 874
Frequency modulation *5:* 921
Fresnel diffraction *4:* 650
Freud, Sigmund *2:* 277
Fried cakes *2:* 356
Friedman, Herbert *6:* 1163
Frisbee *3:* **479-480,** 554
Frisbie, William Russell *3:* 479
Frisch, John G. *3:* 469
Frozen food *3:* 475
Fruit fly *2:* 271, 352, 404; *3:* 546; *4:* 722
Fruits *3:* 569
Frutin, Bernard D. *1:* 20
Fry, Arthur *5:* 856
Fuel *3:* 491, 561
Fuel cell *6:* 990
Fuller, R. Buckminster *3:* 502
Fulton, Robert *6:* 1026
Fungus *6:* 1169
Funk, Casimir *6:* 1137
Furnace *2:* 332; *3:* 532
Fusion bomb *3:* 567

G

Gabor, Dennis *3:* 551
Gagarin, Yuri *6:* 982, 986
Gaia: A New Look at Life on Earth *3:* 481
Gaia hypothesis *3:* **481-482**
Galactosemia *2:* 399
Galaxy *3:* 488, 512; *5:* 868; *6:* 1012, 1114, 1162
Galileo Galilei *2:* 273, 283; *3:* **482-485,** 511, 512, 530, 597; *4:* 647, 710, 738; *5:* 943; *6:* 997, 1032, 1052, 1127
Gallaudet, Thomas *5:* 923
Gall bladder *2:* 348
Gall stones *6:* 1110
Galton, Francis *3:* 461
Galvani, Luigi *2:* 387
Galvanometer *2:* 382; *4:* 649
Game Boy *6:* 1130
Game Gear *6:* 1130
Games *3:* 476, 479; *5:* 907, 935; *6:* 1085, 1129
Game theory *3:* **485-486**
Gamete intrafallopian transfer *3:* 588
Gamma radiation *3:* 527
Gamma ray *2:* 385; *3:* **486-487**; *4:* 645, 747; *6:* 1163
Gamma ray astronomy *3:* **487-488**
Gamma Ray Observatory (GRO) *6:* 1005
Gamow, George *3:* 498
Gardening *3:* 569

Boldfaced numbers indicate main entry pages; italicized numbers indicate volume number

Gas *1:* 19; *2:* 241, 400; *3:* 537, 561; *5:* 841
Gas, existence of *3:* 567
Gas, liquefaction of *3:* 538
Gas mask *2:* 263; *3:* **488-491**; *5:* 927; *6:* 1079
Gasoline *2:* 329, 358; *3:* 471, **491-493,** 562, 586; *4:* 643, 759; *5:* 895
Gasoline engine *3:* 537; *6:* 1018
Gas sylvestre *2:* 241
Gastric physiology *2:* 348
Gastrin *5:* 915
Gas turbine *3:* 595
Gates, William *2:* 295, 312
Gayetty, Joseph C. *6:* 1073
Gayetty's Medicated Paper *6:* 1073
Gay-Lussac, Joseph *3:* 593; *5:* 858, 940
Gay-Lussac's Law *4:* 634
Gay Related Immune Deficiency *1:* 22
Gazzaniga, Michael S. *6:* 1009
Gear *2:* 293
Geber, Jabir ibn Hayyan *1:* 3, 7; *4:* 741
Geiger counter *5:* 873
Gell-Mann, Murray *5:* 862
Gemini *6:* 982, 983, 1119
Gemini 6 *6:* 983
Gemini 10 *6:* 983
Gene *1:* 24; *2:* 269, 275, 352, 404; *3:* **493-495,** 498, 539, 542, 543; *4:* 688, 718, 720, 722; *5:* 919
Genentech *3:* 497
General Electric Company *6:* 1071
General Motors *3:* 492
Generations *2:* 269
Generator *1:* 45
Gene therapy *3:* **495-497**
Genetically engineered blood-clotting factor *3:* **497-498**
Genetic code *3:* **498-499**; *4:* 722
Genetic disorders *3:* 500
Genetic engineering *3:* 497, **499-501,** 539, 541, 542, 553, 583; *4:* 718
Genetic fingerprinting *3:* **501-502**
Genetics *2:* 333, 352
Genetics Institute *3:* 497
Genotype *3:* 495
Geodesic dome *3:* **502-504**
Geography *2:* 319
Geology *2:* 322
Geometric isomers *4:* 708
Geometry *3:* 483; *6:* 1075
Geostationary Operational Environmental Satellites (GOES) *6:* 1151
Geosynchronous orbit *2:* 288
Gericke, W. F. *3:* 569
Germ theory *2:* 237; *3:* **504-508,** 574, 577, 579; *6:* 1064
Germ therapy *2:* 268
Gestation *5:* 859
Giacconi, Riccardo *6:* 1163
Giaever, Ivar *6:* 1099
Giauque, W. F. *2:* 331
Gibbon, John H., Jr. *3:* 529
Gilbert, William *2:* 387
Gin *1:* 38
Giotto *6:* 997
Giraud, John *3:* 463
Glass *1:* 14; *2:* 237, 278, 343, 394; *3:* 456, 457, 476, 587; *4:* 644, 701, 727; *5:* 859, 884, 932; *6:* 1052
Glauber, Johann Rudolf *1:* 7; *3:* 563
Glaucoma *2:* 409; *3:* **508-509**
Glenn, John *6:* 981, 991
Glidden, Carlos *6:* 1102
Global warming *2:* 243; *3:* 482, 513
Globe *6:* 1045
Glucagon *3:* 583
Glucocorticoids *1:* 14
Glucose *5:* 836
Glue *1:* 17
Glycerol *3:* **509**
Goddard, Robert *5:* 901, 902
Godowsky, Leopold *2:* 279
Gold *1:* 7; *2:* 290; *3:* 443, **509-510**; *4:* 684, 741; *5:* 821, 825, 844, 876, 941; *6:* 1088
Gold, Thomas *6:* 1012
Goldmark, Peter *6:* 1058
Gold rush *3:* 510
Golgi, Camillo *4:* 733; *6:* 1039
Gonorrhea *2:* 410; *3:* **510-511**; *6:* 1043
Goodhue, Lyle David *1:* 19
Goodyear, Charles *3:* 444; *5:* 906
Gore, Bob *6:* 1144
Gore-Tex *6:* 1144
Gore, W. L. *6:* 1144
Gorrie, John *1:* 27; *3:* 475
Goudsmit, Samuel *5:* 810
Gould, Charles H. *6:* 1009
Gould, Chester *2:* 284
Governor *2:* 332
Graham, Thomas *2:* 343
Grand Unified Theory *3:* 527
Graphite *2:* 239
Grass *4:* 643
Graves' disease *3:* 552
Graves, Michael *5:* 932
Graves, Robert James *3:* 552
Gravity *2:* 321; *3:* **511-512,** 527; *4:* 724, 738, 739, 752; *5:* 840, 947; *6:* 1010
Gray, Harold *2:* 286
Gray, Stephen *2:* 381
Greatbatch, Wilson *5:* 805
Great pox *6:* 1042
Greek fire *2:* 261
Green, George F. *2:* 340
Greenhouse effect *2:* 243; *3:* **513-516**; *4:* 695; *6:* 1128
Greenhouse gases *3:* 514
Greenwich Mean Time (GMT) *2:* 273
Greenwich Observatory *6:* 1067
Gregor, William *6:* 1070
Gregory, Hanson Crockett *2:* 356
Grenade *3:* **516-517**
Gresser, Ian *3:* 584
Grimaldi, Francesco Maria *5:* 913
Grissom, Virgil "Gus" *6:* 981
Grotthuss-Draper law *5:* 826
Guericke, Otto von *2:* 387
GUIDON *2:* 405
Guillemin, Roger *6:* 1094
Gun silencer *3:* **517-518**
Gunpowder *2:* 360, 396; *3:* 465, 525; *5:* 857, 859, 900
Guthrie, Charles *6:* 1091
Guyot, Arnold Henry *4:* 705
Guyots *2:* 321; *4:* 705
Gymnastics *6:* 1086
Gypsum *3:* 478
Gyrocompass *2:* 292
Gyroplane *3:* 535

H

Haber, Fritz *3:* 456; *4:* 743
Hafnium *2:* 390; *5:* 820
Haggerty, Patrick *2:* 228
Hair care *3:* **519-521**
Hair-care products *2:* 326
Hair dryer *3:* 519
Hair dyes *3:* 520
Hair extensions *3:* 520
Hairpiece *3:* 520
Hair spray *3:* 520
Hair weaving *3:* 521
Haldane, John Burdon *2:* 404; *4:* 722; *5:* 855
Hale, George Ellery *5:* 946; *6:* 1055
Hall, Chester Moor *6:* 1054
Halley, Edmond *4:* 690, 738; *6:* 1148
Hall, Lloyd A. *3:* 476
Hall, Samuel Read *6:* 1045
Hallucinogen *3:* **521-523**
Halogen lamp *2:* 394; *3:* **523-524**
Halol *6:* 1088
Haloperidol *6:* 1088
Halsted, William *2:* 277
Haltran *3:* 572
Hamilton, Alice *4:* 741
Hammond, Laurens *4:* 720
Hancock, Thomas *5:* 905
Handguns *2:* 273
Hanway, Jonas *6:* 1115
Hard disk *2:* 303
Hard drive *2:* 305
Harder, Delmar S. *2:* 333
Harding, Warren G. *5:* 875
Hardy, Godfrey Harold *5:* 855
Hardy-Weinberg equilibrium *5:* 855
Harington, Sir John *6:* 1072
Harmine *3:* 522
Harpoon *3:* **524-525**
Harrington, George F. *2:* 340
Harrington, Joseph *2:* 307
Harris, Geoffrey *3:* 553
Harris, John *6:* 1074
Harrison, Michael *5:* 860
Harrison, Ross Granville *4:* 734
Harris, Rollin *2:* 293
Hartley, Walter Noel *4:* 772
Hart, William Aaron *4:* 723
Harvard Graphics *2:* 297
Hashish *3:* 522
Hata, Sahachiro *6:* 1043
Hausdorff, Felix *6:* 1076
Hawking radiation *3:* 527
Hawking, Stephen William *3:* **526-527**
Healing *1:* 15
Health *2:* 256, 263, 269, 339, 343; *3:* 447, 468, 471, 478, 504; *4:* 673, 682, 687; *5:* 805
Hearing *1:* 12
Hearing aids and implants *3:* **528-529**; *4:* 697
Hearing impairment *2:* 315; *5:* 921
Heart *2:* 256
Heart defects, congenital *3:* 530
Heart disease *2:* 269, 382
Heart-lung machine *3:* **529-530**
Heat *1:* 27; *2:* 385; *3:* 526; *4:* 688
Heat and thermodynamics *1:* 3; *3:* **530-532**; *5:* 924
Heating *3:* **532-535**
Heat pump *3:* 535
Heat-resistant glass *6:* 1055
Heezen, Bruce Charles *4:* 704
Heezen-Ewing theory *4:* 704
Heinlein, Robert A. *6:* 1143
Heisenberg, Werner *5:* 865, 866, 889
Helicopter *1:* 33, 36; *3:* 503, **535-537**; *6:* 1078
Heliocentric theory *5:* 838
Heliography *5:* 832
Helium *2:* 330; *3:* **537-539,** 599; *4:* 637, 726; *5:* 866, 883; *6:* 1033, 1159
Helmont, Jan Baptista van *3:* 569
Hemley, R. J. *3:* 566
Hemophilia *3:* 497, 539-541; *5:* 920
Hench, Philip *1:* 14; *3:* 552
Henie, Sonja *3:* 573
Henry, Edward R. *3:* 461
Henry, Prince of Portugal *2:* 291
Hepatitis *1:* 15; *3:* **541-542,** 584
Heredity *2:* 269, 333, 404; *3:* 493, 498, 499, 522, 539, **543-547**; *4:* 688
Heroin *3:* **547**; *4:* 713
Hero of Alexandria *4:* 653; *5:* 900
Herpes *2:* 263; *3:* 542, 584
Herschel, William *3:* 461; *4:* 677, 728; *5:* 913; *6:* 1054, 1118, 1119
Hershey Chocolate Company *2:* 267
Hershey, Milton S. *2:* 267
Hertz *1:* 45; *2:* 385
Hertz, Heinrich Rudolph *2:* 388; *3:* 590; *4:* 655; *5:* 829
Hess, Harry Hammond *2:* 321; *4:* 705
Hess, Victor Franz *6:* 1163
Hevelius, Johann *4:* 710
Hevesy, György *3:* 594
Hewish, Antony *5:* 863
Hieroglyphics *1:* 41
Higgs particle *5:* 810
High-definition television (HDTV) *2:* 227; *3:* 459; *6:* 1060
High Energy Astrophysical Observatories (HEAO) *3:* 488; *6:* 1005
High-pressure physics *3:* **547-548**
High risk behavior *1:* 20
High-speed flash photography *3:* **548-549**
Highways *2:* 329
Hildebrand, Alan *2:* 329
Hilyer, Andrew Franklin *3:* 535
Hindenburg *1:* 33
Hindu-Arabic numerals *5:* 905
Hinton, William A. *6:* 1043
Hippocrates *4:* 717
Histamine *1:* 39
Histology *6:* 1069, 1070
Hitchings, George *2:* 263
Hitler, Adolf *1:* 6
H.L. *Hunley* *6:* 1078
H.M.S. *Challenger* *4:* 757
Hockey *3:* 572
Hodge, P. R. *3:* 464
Hodgkin, Dorothy Crowfoot *5:* 817; *6:* 1162

Boldfaced numbers indicate main entry pages; italicized numbers indicate volume number

Hodgkin's disease *4:* 662
Hoffman, Albert *3:* 523
Hoffmann, Erich *6:* 1043
Hofstadter, Robert *4:* 735; *5:* 862
Hog ring machine *6:* 1010
Holland, John *6:* 1026
Hollerith, Herman *2:* 300
Hologram *3:* **549-551**
Home Box Office (HBO) *2:* 225
Homeostasis *2:* 400; *3:* 552
Homo erectus *3:* 556
Homo habilis *3:* 555
Homo sapiens *3:* 556
Homo sapiens neanderthalensis *3:* 557
Homo sapiens sapiens *3:* 554
Hooke's law *2:* 379; *5:* 935
Hopper, Grace *4:* 666
Hormone *1:* 14; *2:* 263, 269, 322, 399; *3:* **552-553,** 561, 581; *5:* 914; *6:* 1094
Hornbook *6:* 1045
Hot combs *3:* 520
Housing *3:* 502
Houssay, Bernardo *3:* 553
Houston, Tom *6:* 1152
Hoyle, Fred *6:* 1013
HP-35 *2:* 231
Hubble, Edwin Powell *5:* 887; *6:* 1012
Hubble Space Telescope (HST) *3:* 599, 601; *6:* 1005
Hughes, Howard *1:* 33
Hula hoop *3:* **554**
Human evolution *3:* **554-560**
Human gene therapy *3:* 500
Human growth hormone (somatotropin) *3:* 553, **560-561**
Human immunodeficiency virus (HIV) *1:* 20, 24; *2:* 263; *5:* 893
The Human Use of Human Beings: Cybernetics and Society *2:* 333
Humidity *1:* 27
Humulin *3:* 583
Hunter, John *6:* 1043
Hunter's chancre *6:* 1042
Hunting *3:* 524
Huntington disease *4:* 731
Huntington's chorea *3:* 500
Hurricanes *6:* 1020
Hutchinson, Miller Reese *3:* 528
Hutton, James *3:* 481; *6:* 1116
Huxley, Julian *2:* 404
Huygens, Christiaan *2:* 273, 396; *4:* 677; *5:* 913
Hyatt, John Wesley *5:* 842
Hyde, James F. *5:* 925
Hydrocarbon *2:* 401; *3:* **561-562,** 566; *4:* 725; *5:* 914
Hydrochloric acid *1:* 9; *2:* 347; *3:* **562-563,** 565; *4:* 741, 766; *5:* 914
Hydrochlorofluorocarbon *1:* 20
Hydrofluoric acid *3:* 470
Hydrofluorocarbon *1:* 20
Hydrofoil *3:* **563-564**
Hydrogen *2:* 283, 330; *3:* 538, 561, 563, **564-567,** 599; *4:* 751; *5:* 836, 858, 912; *6:* 1032, 1124, 1170
Hydrogenation *4:* 760
Hydrogen bomb *3:* **567-568**
Hydroponics *3:* **569**
Hydrostatic balance *3:* 483
Hygrometer *6:* 1148
Hyperglycemia *3:* 561
Hyperopia *2:* 407, 410
Hypertension *3:* **570**
Hypoglycemia *3:* 561
Hypothalamus *6:* 1094

I

IBM Corporation *2:* 300, 316; *3:* 530; *4:* 666; *6:* 1104
Ibuprin *3:* 572
Ibuprofen *3:* **571-572**
Icarus *5:* 912
Ice Capades *3:* 573
Ice cream *6:* 1174
Ice-making machine *3:* 475
Ice-resurfacing machine *3:* **572-573**
Ice skating *3:* 572
Iconoscope *6:* 1058
Ideal Toy Company *5:* 907
I G Farben *2:* 412
Igneous rock *6:* 1070, 1071
Ignition system *2:* 397
Illness *3:* 508
Illustrations of the Huttonian Theory *6:* 1117
Immune system *1:* 22, 24, 39; *3:* **573-575,** 579; *5:* 824, 848; *6:* 1064, 1173
Immunology *3:* 573
Immunosuppression *6:* 1093
Inapsine *6:* 1088
Incas *3:* 510
Incubation *6:* 1063
Incubation period *3:* 542
Incus *1:* 13
Industry *2:* 282, 384, 385; *3:* 561, 564, 591
Inertia *3:* 511
Infantile glaucoma *3:* 508
Infantile paralysis *5:* 849
Infectious hepatitis *3:* 542
Inflationary theory *6:* 1013
Influenza *5:* 848
Infrared *2:* 280
Infrared Astronomical Satellite (IRAS) *6:* 1005
Infrared light *5:* 891
Infrared radiation *2:* 282; *3:* 515; *6:* 1005, 1163
Injection molding *5:* 854
Ink *1:* 14; *4:* 764
Ink jet printer *2:* 310
Inoculation *3:* 504, 573, **575-579**; *5:* 849; *6:* 1043, 1064
Inquisition *3:* 482
Insecticide *2:* 336
Instant camera (Polaroid Land camera) *2:* 279; *3:* **579-580**; *5:* 834
Instant coffee *3:* 475, **580-581**
Insulating technique *5:* 906
Insulation *3:* 456
Insulin *3:* 499, 552, **581-583**; *6:* 1007, 1162, 1166
Integrated circuit *2:* 228, 300, 302; *4:* 697
The Integrative Action of the Nervous System *6:* 1040
Intensity ratings *2:* 365
Interferometer *5:* 881
Interferon *1:* 25; *3:* **583-584**
Internal combustion engine *2:* 396; *3:* 464, 493, **584-587,** 594
International Business Machines *2:* 302
International Date Line *6:* 1067
International Ultraviolet Explorer (IUE) *6:* 1003
Intestine *5:* 914

Inventions or Devices *6:* 1024
In vitro fertilization *3:* **587-588**
Iodine *5:* 876
Iodine-131 *5:* 877
Ion *1:* 8; *4:* 708; *5:* 841, 949
Ionosphere *2:* 370; *3:* **588-591**; *5:* 921; *6:* 1035
Ipatiev, Vladimir *4:* 694
Iridium *5:* 845
Iron *2:* 245, 368, 386; *3:* 565, **591-593**; *4:* 667, 674, 676, 712, 771; *5:* 858, 909; *6:* 1068, 1088
Iron lung *5:* 849
Irradiation *3:* 476
Irrational number *3:* **593**; *5:* 883
Irrigation *2:* 231
Isaacs, Alick *3:* 583
Isla, Rodrigo Ruiz de *6:* 1042
Isobar maps *6:* 1148
Isomer *2:* 239; *3:* **593-594**; *4:* 707
Isotope *3:* 566, **594**; *5:* 858, 876, 940; *6:* 1090, 1118, 1160
Israel *2:* 377
Ivanovsky, Dmitri *6:* 1134
Ives, Frederic Eugene *2:* 279

J

Jackup rigs *4:* 763
James I *6:* 1024
James, Richard *5:* 935
Japan Victor Company (JVC) *6:* 1132
Jaundice *5:* 892; *6:* 1172
Java man *3:* 558
Jefferson, Thomas *3:* 577
Jeffreys, Alec *3:* 502
Jenner, Edward *3:* 573, 577
Jenney, William Le Baron *5:* 931
Jennings, Harry C. *6:* 1151
Jennings, Thomas L. *2:* 359
Jervis, John B. *6:* 1083
Jet engine *1:* 34; *2:* 396; *3:* 587, **595-597**; *6:* 1014
Jet pack *6:* 993
Jet stream *3:* 593, 594; *6:* 1149
Jewelry *3:* 509
Johannsen, Wilhelm *2:* 271, 404; *3:* 495, 546
Johanson, Donald C. *3:* 559
John VI *3:* 529
John Bull *6:* 1080
Johnson, Reynold *2:* 303
Jolly, Jean-Baptiste *2:* 358
Josephson, Brian *6:* 1100
Joule, James Prescott *1:* 2
Judson, Whitcomb L. *6:* 1176
Junk food *5:* 938
Jupiter *3:* 484, 566, 599, **597-601**; *4:* 694, 729, 738; *5:* 846, 880, 912, 943, 946, 949; *6:* 995, 1006, 1118

K

Kahanamoku, Duke Paoa *6:* 1037
Kahn, Julian S. *1:* 19
Kahn, Philippe *2:* 295
Kahn, Reuben Leon *6:* 1043
Kaleidoscope *4:* **633**
Kalium *5:* 857
Kamerlingh, Onnes Heike *2:* 330; *3:* 538; *5:* 809; *6:* 1035
Kant, Immanuel *6:* 1033
Kapany, Narinder S. *3:* 457
Kaposi's sarcoma *1:* 22
Katz, Bernard *1:* 6
Kekulé, Friedrich *4:* 707; *6:* 1125
Kelly, Walt *2:* 286
Kelly, William *6:* 1093
Kelsey, Frances *6:* 1062
Kelvin, Lord *2:* 292, 388
Kelvin scale *1:* 2
Kempelen, Wolfgang von *6:* 1138
Kendall, Edwin Calvin *1:* 14; *3:* 520
Kennedy, John F. *6:* 981
Kennelly-Heaviside layer *3:* 588
Kent, Arnold *1:* 27
Kepler, Johannes *3:* 484; *5:* 838, 840; *6:* 1052
Keratitis *2:* 409
Keratoconus *2:* 319
Keratometer *2:* 319
Kerogen *5:* 822
Kerosene *2:* 358, 396; *3:* 491; *4:* 760
Kevlar *3:* 460; *4:* **633-634**
Kidneys *2:* 343
Kieselguhr *2:* 360
Kilby, Jack *2:* 229
Kimberlite *2:* 372
Kinescope *6:* 1058
Kinetic energy *4:* 748
Kinetic theory of gases *4:* **634-635**
Kinetograph *4:* 716
Kinetoscope *4:* 714, 716
Kipping, Frederic Stanley *5:* 924
Kirchhoff, Gustav *6:* 1006, 1032
Kitasato, Shibasaburo *6:* 1063, 1064
Kite *1:* 29; *4:* **635-636**; *6:* 1147
Klaproth, Martin Heinrich *6:* 1070, 1117
Klein, Christian Felix *6:* 1076
Knerr, Rich *3:* 480
Knight, Gowin *2:* 292
Knight, J. P. *6:* 1079
Knowledge base *2:* 405
Koch, Robert *2:* 268; *3:* 505, 578; *6:* 1064
Kodachrome *2:* 279
Kolbe, Adolf Wilhelm Hermann *1:* 4
Koller, Carl *2:* 277
Kölliker, Rudolf Albert von *4:* 733
Kolorev, Sergei *6:* 985
Komarov, Vladimir *6:* 986
Krumm, Charles L. *6:* 1051
Krypton *3:* 539; *4:* **637-638,** 726; *5:* 883; *6:* 1159
Kuhn-Moos, Catharina *6:* 1176
Kuiper, Gerard *5:* 846, 914
Kurchatov, Igor *3:* 594
Kurzweil music synthesizer *6:* 1041

L

Lacquer *1:* 14
Lactic acid *3:* 508

Boldfaced numbers indicate main entry pages; italicized numbers indicate volume number

Lake Nyos *2:* 243
Lamarck, Jean Baptiste de *2:* 402; *3:* 544, 557
Lamy, C. A. *6:* 1065
Land, Edwin Herbert *2:* 279; *3:* 579; *4:* 651; *5:* 834; *6:* 1066
Land bridge *2:* 319
Land masses *2:* 319
Landsats *2:* 373
Landsteiner, Karl *5:* 850, 892
Lane, Hugh Arbuthnot *3:* 479
Langmuir, Irving *5:* 841
Language, universal *4:* **639**
Lanier, Jaron *2:* 314
Lanolin *6:* 1143
Lanthanum *5:* 820
Laplace, Pierre *6:* 1033
Laptop computers *2:* 305; *4:* 695
Lard *3:* 447
Laser *2:* 255, 290; *3:* 551; *4:* **640-642,** 764, 765; *5:* 834, 872, 942; *6:* 1157
Laser chemistry *5:* 827
Laser printer *2:* 310
Lassone, Joseph Marie François de *2:* 245
Latex *5:* 905
Lathe *2:* 305
Latin *4:* 639
Latour, Charles Cagniard de *6:* 1169
Latrines *6:* 1072
Laue, Max von *6:* 1162, 1166
Laundromat *4:* **642**
Laundry *2:* 276
Laurent, Alexander *3:* 462
Lavoisier, Antoine-Laurent *2:* 283; *4:* 768
Lawn mower *4:* **643-644**
Lawn, Richard *3:* 498
Law of conservation of energy *3:* 531
Law of constant composition *6:* 1124
Law of entropy *3:* 531
Law of Transmission by Contact *4:* 734
Law of uniformly accelerated motion *3:* 483
Laws of motion *4:* 738
LCD (liquid crystal display) *4:* **644,** 651
Lead *4:* 645, 723, 766; *5:* 928; *6:* 1063, 1088
Leakey, Louis S. B. *3:* 559
Leakey, Mary Douglas *3:* 559
Leather *2:* 410
LED (light-emitting diode) *2:* 230; *4:* 644, **645-647,** 728
Leeuwenhoek, Antoni van *4:* 647
Legionnaires' disease *5:* 848
Lehmann, Inge *5:* 918
Leibniz, Gottfried *4:* 738
Leith, Emmet *3:* 551
Lemelson, Jerome *5:* 898
Lemoine, Louis *5:* 895
Lenard, Philipp *5:* 829
Lenoir, Jean-Joseph Étienne *2:* 396; *3:* 584
Lens *2:* 318; *3:* 467; *4:* **647-648**
Lenz, Widukind *6:* 1062
Leonardo da Vinci *1:* 27, 29; *3:* 537, 556; *6:* 1148
Leonov, Alexei *6:* 986
Leptons *5:* 810; *6:* 1024
Lesch-Nyhan syndrome *3:* 500
Leukemia *2:* 410; *3:* 497; *6:* 1108
Leupold, Jacob *6:* 1017
Lever *6:* 1016
Leverrier, Urbain *5:* 840
Levi-Montalcini, Rita *4:* 730
Levitt, Arnold *2:* 357
Lewis, Gilbert Newton *1:* 8
Lewis, John *2:* 340
Lexell, Anders Johan *6:* 1119
Leyden jar *2:* 382
Libration *4:* 710
Lie detector *4:* **649-650**
Liebig, Justus von *1:* 8; *5:* 858
Light *1:* 44; *2:* 380; *3:* 456, 467, 512; *4:* 633, 640, 645, 694
Light bulb, incandescent *3:* 467, 523
Light, diffraction of *2:* 282; *4:* 648, **650,** 656; *6:* 1007
Lighthouse *2:* 301
Lightning rod *2:* 382
Light, polarization of *4:* **650-651**
Light ray *4:* 653; *6:* 1052
Light, reflection and refraction of *4:* 648, **651-653,** 656
Light, theories of *2:* 282; *4:* 651, 652, **653-656,** 739; *6:* 1145
Light waves *2:* 385
Lillehei, Richard *6:* 1093
Limestone deposition *3:* 515
Lindbergh, Charles *1:* 33; *6:* 981
Linde, Carl von *4:* 743
Lindenmann, Jean *3:* 583
Link trainer *2:* 313
Linnaeus, Carolus *3:* 557
Lipid *1:* 4; *4:* **656-657**
Lipman, Hyman *2:* 400
Lippershey, Hans *3:* 484; *6:* 1052
Lippmann, Gabriel *2:* 279
Lip reading *5:* 923
Liquid gas *6:* 1123
Liquid Paper *6:* 1104
Lister, Joseph *3:* 479, 508, 577
Listing, Johann Benedict *6:* 1075
Lithium *5:* 858
Lithium nucleus *4:* 748
Little, A. B. *3:* 464
Liver cancer *3:* 541
Liver disease *3:* 541
Local anesthetics *4:* 747
Lock *5:* 909
Lock and key *4:* 635, **657-659**; *5:* 910
Lockjaw *6:* 1064
Locomotive *2:* 301, 398; *6:* 1029
Lodestone *2:* 290
Loewi, Otto *1:* 6
Löffler, Friedrich August *6:* 1135
Logarithm *2:* 339; *4:* **660**
Long playing (LP) records *4:* 669
Long radio waves *2:* 280
Loom *2:* 250
L'Oréal laboratories *3:* 520
Lotus Development Corp. *2:* 295
Lotus Notes *2:* 297
Loud, John J. *5:* 815
Loudspeaker *6:* 1019
Louis XIV *3:* 465
Lovelock, James *3:* 481
Lower, Richard *6:* 1092
LSD *3:* 522
Lucite *1:* 13

Lucy *3:* 559
Lumière, Auguste *2:* 279; *4:* 715, 716
Lumière, Louis *2:* 279; *4:* 715, 716
Luna 1 *6:* 994
Luna 2 *4:* 711
Luna 3 *4:* 711; *6:* 994
Luna 9 *6:* 994
Lunar eclipses *4:* 709
Lunar topography *4:* 710
Lung cancer *3:* 497
Lungs *3:* 529
Lunokhod 1 *4:* 712
Luvs *2:* 345
Lyell, Charles *2:* 403; *6:* 1117
Lyme disease *4:* **661**
Lymphatic system *4:* **662-663**
Lymph glands *4:* 662
Lymphocytes *4:* 662
Lyot, Bernard *5:* 947

M

McAdam, John *5:* 895
McCollum, Elmer V. *6:* 1137
McCormick, Cyrus *4:* 680
Machine language *4:* **665-666**
Machine tools *4:* 679
Machine vision *2:* 317
Macintosh, Charles *6:* 1143
McKay, Frederick S. *3:* 468
McMillan, Edwin *4:* 730
Machine gun *1:* 31
Magellan *6:* 997, 1129
Magic Tape *1:* 19
MAGLEV technology *6:* 1084
Magnesium *1:* 10; *4:* **667-668,** 712
Magnet *2:* 382; *5:* 809
Magnetic core memory *4:* 672
Magnetic declination *2:* 291, 370
Magnetic dip *2:* 370, 387
Magnetic field *2:* 331; *3:* 487, 599; *4:* **668,** 752; *5:* 921, 948; *6:* 1035, 1108, 1128
Magnetic levitation *6:* 1084
Magnetic recording *2:* 303; *4:* **668-670,** 697, 699, 745; *6:* 1019, 1132
Magnetic resonance *4:* 753
Magnetic resonance imaging (MRI) *4:* **670-671,** 753
Magnetic tape *2:* 308; *4:* 668
Magnetism *2:* 291, 385, 387
Magnetometer *4:* 757
Magneto-optical shutter *3:* 548
Magnetophon *4:* 669
Magnifying glass *6:* 1038
Magnitude *2:* 366
Mahoney, John F. *6:* 1044
Maiman, Theodore Harold *2:* 290; *3:* 551; *4:* 641
Mainframe *2:* 298
Mainframe computer *2:* 295, 310, 312; *4:* **671-673,** 705
Makeup *2:* 322
Malaise *3:* 541
Malaria *2:* 336; *6:* 1172
Malleus *1:* 13
Malpighi, Marcello *4:* 662
Mammals *2:* 326
Mammography *4:* **673-674**
Manby, George William *3:* 462, 464
Manganese *4:* **674-676**; *6:* 1048
Manhattan Project *2:* 376; *3:* 567; *4:* 748, 753
Mannes, Leopold *2:* 279
Manufacturing *2:* 282, 305, 317; *3:* 562
Manuscripts *2:* 400
Mao, H. K. *3:* 566
Map *2:* 291; *3:* 449
Marcel waves *3:* 519
Marconi, Guglielmo *3:* 590; *4:* 635; *5:* 875, 920; *6:* 1056
Marey, Etienne-Jules *4:* 716
Margarine *3:* 448
Maria *4:* 710
Marie-Davy, Edme Hippolyte *6:* 1148
Marijuana *3:* 522
Mariner 2 *5:* 948; *6:* 1128
Mariner 9 *5:* 912
Mariner 10 *4:* 687; *5:* 942
Mark I *2:* 300, 302
Mars *2:* 242; *4:* **676-678,** 729; *5:* 911, 912, 945; *6:* 995, 1118
Marum, Martin van *4:* 772
Maser *4:* 702, 765
Mash *1:* 37
Masking tape *1:* 18
Mason, Walter *2:* 342
Mass production *2:* 252, 326, 333, 394; *4:* **678-681**
Mass spectrograph *4:* **681-682**
Mass spectrometer *4:* 681
Mass transit *6:* 1028
Math *1:* 38; *2:* 228, 293; *3:* 483, 485, 593; *5:* 855, 883; *6:* 1075
Mathematical modeling *6:* 1147
Mathijsen, Antonius *3:* 478
Matrix mechanics *5:* 865
Matrix theory *5:* 868
Matsushita *6:* 1132
Matter *3:* 512
Mauchly, John *2:* 303
Maxim, Hiram *3:* 519
Maxim silencer *3:* 517
Maxwell, James Clerk *2:* 388; *4:* 655; *5:* 867
Maya *2:* 264, 266
Maybach, Wilhelm *3:* 586
Mayer, Tobias *4:* 710
Maytag Company *6:* 1142
Mazzitello, Joe *6:* 1131
Me 262 *3:* 595
Measles *3:* 542, 575; *6:* 1134
Medawar, Peter *3:* 575; *6:* 1092
Medical diagnostic devices *4:* 674
Medical technology *2:* 254, 256, 263, 343, 382, 383, 393, 405; *3:* 456, 468, 478, 495, 497, 499, 528, 529, 575, 581, 583, 587; *4:* 640, 670, 674, 713, 747, 751; *5:* 805, 806, 809, 849, 859, 872, 881, 891, 908; *6:* 1030, 1041, 1090, 1108, 1151
Medicine *2:* 322, 343; *3:* 478, 549, 571, 583; *4:* 640
Medipren *3:* 572
Meiosis *3:* 494
Melanoma *3:* 496
Melin, Spud *3:* 480
Mendel, Gregor *2:* 271, 352, 404; *3:* 495, 545
Mendeleev, Dmitry I. *5:* 820, 821

Boldfaced numbers indicate main entry pages; italicized numbers indicate volume number

Mendez, Arnaldo Tomayo *6:* 982
Meningitis *4:* **682-683**; *5:* 818
Mercalli, Guiseppe *2:* 365
Mercury (element) *4:* **684-686,** 768; *6:* 1042, 1114
Mercury (planet) *4:* **686-687**; *5:* 945; *6:* 995, 1118
Mercury poisoning *4:* 685
Merryman, Jerry *2:* 230
Mescaline *3:* 522
Mestral, Georges de *6:* 1126
Metabolism *2:* 322, 398; *3:* 560, 583, 593; *4:* **687-688**; *6:* 1137
Metal fatigue *4:* **688-689**
Metals *1:* 40
Metchnikoff, Elie *3:* 574; *5:* 824; *6:* 1174
Meteor and meteorite *3:* 593; *4:* **689-692**
Meteorite *2:* 240; *4:* 687, 726; *6:* 1070
Meteorology *4:* **692-694**; *6:* 1148
Methadone *4:* 713
Methane *3:* 514, 562; *4:* **694-695,** 725; *5:* 910, 914; *6:* 1120
Methodius, Saint *1:* 44
Metzger, Ray *6:* 1152
Meyer, Julius Lothar *5:* 820
Michel, Helen *2:* 327
Michell, John *5:* 917
Microbes *3:* 505
Microbiology *3:* 508
Microchip *4:* 669
Microcomputer *2:* 298, 304, 310, 312, 406; *4:* **695-697**; *5:* 900; *6:* 1035, 1130
Microfiche *2:* 310
Microfilm *5:* 827
Micrometer *2:* 406; *5:* 909
Microphone *2:* 316, 348; *3:* 528; *4:* 696, **697-699**; *5:* 874; *6:* 1019
Microprocessor *2:* 300; *4:* 697, **699**; *5:* 899
Microscope *6:* 1151
Microscopy *4:* 733
Microsoft Corporation *2:* 295
Microtone *3:* 528
Microwave *4:* **700**
Microwave oven *4:* 695, 699, **700-702**
Microwave transmission *4:* **702-704**
Midatlantic Ridge *4:* 704
Midgley, Thomas, Jr. *1:* 28; *3:* 492
Mid-ocean ridges and rifts *2:* 321; *4:* **704-705,** 758; *5:* 843
Midol 200 *3:* 572
Mig 17 *3:* 595
Military equipment *3:* 465, 466
Milk-Bone *2:* 356
Milk chocolate *2:* 266
Milky Way *3:* 484, 488; *6:* 1010, 1011, 1164
Mill *4:* 679
Mill, Henry *6:* 1101
Miller, Carl *5:* 828
Milling machine *2:* 305
Mills, H. S. *5:* 936
Miner's Friend *6:* 1017
Mini-black hole *3:* 527
Minicomputer *2:* 298, 310, 312; *4:* **705**; *6:* 1035
Mining *2:* 359; *3:* 488
Minnesota Mining and Manufacturing Company (3M) *1:* 18; *5:* 856; *6:* 1131, 1144
Mir *6:* 987, 1001
Miranda *6:* 1119
Mirror *2:* 278, 290; *6:* 1052
MITalk System *6:* 1139
Mitosis *2:* 270
Mnemonics *4:* 666
Mobile station *5:* 876
Möbius, August Ferdinand *6:* 1076
Model T *4:* 680
Modem *2:* 308
Modified Mercalli Scale *2:* 365
Mohorovicic, Andrija *2:* 370, 371; *4:* 706; *5:* 918
Mohorovicic discontinuity *2:* 370, 371; *4:* **705-707**; *5:* 918
Moivre, Abraham de *3:* 486
Mold *3:* 472
Molecular biology *4:* 709
Molecular structure *3:* 447, 593; *4:* 651, **707-709**; *5:* 827; *6:* 1162
Molecular weight *1:* 105-107
Monarch *6:* 1050
Money *2:* 252
Monorail system *6:* 1085
Monsanto *3:* 448
Monsoons *6:* 1021
Montagu, Mary Wortley *3:* 573, 575
Montgolfier, Jacques-Étienne *1:* 29
Montgolfier, Joseph-Michel *1:* 29
Moog, Robert *4:* 719; *6:* 1041
Moon *3:* 484, 512; *4:* 677, **709-712**; *5:* 826, 914; *6:* 994
Morgan, Garrett Augustus *3:* 488; *6:* 1079
Morgan, Thomas Hunt *2:* 271, 404; *3:* 495, 546; *4:* 722
Morita, Akio *6:* 1141
Morphine *2:* 277, 393; *3:* 547; *4:* **713**
Morrison, James B. *2:* 340
Morrison, Walter Frederick *3:* 480
Morse code *5:* 920; *6:* 1051
Morveau, L. B. Guyton de *2:* 283
Moseley, Henry *5:* 821; *6:* 1048
Mosquito *2:* 336; *6:* 1171
Mother of pearl *3:* 444
Motion picture *2:* 325; *3:* 551; *4:* **713-715, 717**; *5:* 829; *6:* 1046, 1060
Motorcycle *3:* 585
Motrin *3:* 571
Motz, Lloyd *5:* 889
Mouse *2:* 309
Mouth *2:* 346
Mouthwash *3:* 469
Movie. *See* Motion picture
Movie camera *4:* **715-717**
M13 *6:* 1010
Muller, Hermann *2:* 271, 404; *4:* 722
Müller, K. Alex *6:* 1036
Multiple sclerosis *1:* 15
Multiplication *1:* 38; *2:* 227, 228
Multiplication tables *4:* 660
Multistage flash distillation *2:* 343
Multiwire Proportional Counter (MPC) *5:* 873
Mumps *3:* 542; *4:* **717**; *6:* 1135

Muscular dystrophy *4:* **717-718**; *5:* 920
Music *4:* 668
Musical instrument *5:* 897
Musical instrument, electric *4:* **718-720**
Musket *4:* 679
Mustard gas *2:* 262
Mutagen *4:* 720
Mutagenesis *4:* 720
Mutation *3:* 545; *4:* **720-722**
Muybridge, Eadweard *4:* 713, 716
MYCIN *2:* 405
Myers, Ronald *6:* 1008
Mylar *6:* 991
Myopia *2:* 407, 410; *5:* 872

N

Nabisco *2:* 356
Napier, John *2:* 339; *4:* 660
Napoleon *6:* 1172
National Advisory Committee for Aeronautics (NACA) *6:* 1156
National Aeronautics and Space Administration (NASA) *3:* 488; *5:* 941; *6:* 981, 1005
National Cancer Institute *4:* 674
National Cash Register Company *2:* 254
National Institute of Drycleaning *2:* 358
National Institutes of Health Recombinant DNA Advisory Committee *3:* 496
National Manufacturing Company *2:* 254
National Radio Astronomy Observatory (NRAO) *5:* 881
National Television System Committee (NTSC) *6:* 1060
Native Americans *5:* 922
Natta, Giulio *5:* 843
Natural gas *2:* 238; *3:* 561; *4:* 694, **723-725,** 726; *5:* 822
Natural selection *2:* 403; *3:* 544; *4:* 722; *5:* 855
The Nature of the Chemical Bond *5:* 811
Nautilus *2:* 292; *6:* 1026
Navigation *2:* 275, 291
Navigational satellite *4:* 704, **725-726**; *6:* 1055
Neanderthals *3:* 558
Nearsightedness *2:* 407, 410
Nebula *5:* 868; *6:* 1012, 1033, 1112
Nei Ching *1:* 16
Neisser, Albert *6:* 1043
Nelmes, Sarah *3:* 577
Neon *3:* 538; *4:* **726-727**; *5:* 883; *6:* 1033, 1159
Neon light *4:* **727-728**
Neopangaea *5:* 843
Neosalvarsan *6:* 1044
Neptune *4:* 695, **728-729,** 730; *5:* 845, 946
Neptunium *4:* **729-730**
Nero *2:* 410
Nerve fibers *4:* 733; *6:* 1039
Nerve growth factor *4:* **730-731**
Nervous system *1:* 4; *4:* 685, **731-733**; *5:* 813, 849, 914, 939; *6:* 1064
Nesmith, Bette Graham *6:* 1104
Nestlé *2:* 267
NetWare *2:* 297
Neumann, John von *3:* 485; *4:* 665
Neurology *1:* 6; *4:* 718
Neuron theory *1:* 4; *3:* 570; *4:* **733-734**; *6:* 1039
Neurotransmitter *1:* 5
Neutron *3:* 566; *4:* **734-735**; *5:* 862, 863; *6:* 1118, 1022, 1090
Neutron bomb *4:* **736-737**
Neutron star *5:* 863, 879; *6:* 1164
Newcomb, Simon *5:* 845
Newcomen, Thomas *6:* 1017
New England Digital Synclavier *6:* 1041
Newland, J. A. R. *5:* 819
Newton, Isaac *2:* 281, 283, 375; *3:* 512, 526; *4:* 647, 653, **737-739**; *5:* 840, 867; *6:* 1006, 1032, 1054
Niacin *4:* **739-740**
Nichols, Larry D. *5:* 907
Nicholson, Seth Barnes *6:* 1128
Nickel *2:* 245, 368; *4:* 692
Niemann, Albert *2:* 277
Niépce, Joseph Nicéphore *5:* 831
Nightingale, Florence *3:* 577
Niña, La *4:* 693; *6:* 1148
Niño, El *4:* 693
Nintendo *6:* 1130
Niobium *6:* 1068
Nipkow, Paul *6:* 1056
Nippon Electric Corporation *2:* 316
Nissen, George *6:* 1085
Nitric acid *1:* 7; *4:* **740-741,** 744, 766; *5:* 845
Nitrogen *2:* 330, 339; *3:* 514, 590; *4:* 637, 656, **741-744,** 758; *5:* 887, 914, 949; *6:* 1033, 1090
Nitrogen oxide *1:* 10
Nitroglycerin *2:* 359
Nitrous oxide *3:* 445; *4:* 744
Nobel, Alfred *2:* 359
Nobel, Ludwig *2:* 249
Nobel Prizes *1:* 6; *2:* 359, 375; *5:* 813; *6:* 1094, 1100
Nobel's Safety Powder *2:* 361
Noble gases *5:* 820
Noise reduction system *4:* **744-745**; *6:* 1019
Nollette, Abbé Jean Antoine *4:* 766
No More War! *5:* 812
Non-sinusoidal *4:* 765
Non-verbal communication *5:* 921
Norman, Robert *2:* 292
Northern Lights *3:* 588
Nova *5:* 863
Nova and supernova *4:* **745-747**
Novell Inc. *2:* 295
Novocain *2:* 278; *4:* **747**
Nowcasting *6:* 1146
Nuclear fission *2:* 376, 377; *4:* 730, **748-749,** 751; *6:* 1118
Nuclear fusion *3:* 567; *4:* **749-751**; *5:* 841; *6:* 1032
Nuclear magnetic resonance (NMR) *4:* **751-753**

Boldfaced numbers indicate main entry pages; italicized numbers indicate volume number

Nuclear physics *5:* 877
Nuclear radiation *3:* 486
Nuclear reactor *4:* **753-755**
Nucleic acid *2:* 353; *3:* 471; *4:* 656, 709; *5:* 912; *6:* 1133
Nucleoplasm *2:* 260
Nucleotide *3:* 498
Nucleus *2:* 258; *4:* 731; *5:* 841
NutraSweet *3:* 448
Nutriculture *3:* 569
Nutrition *3:* 447
Nylon *2:* 252, 335; *3:* 460; *4:* **755-756**; *5:* 843, 851, 853, 855; *6:* 1074, 1115, 1126

O

Oberon *6:* 1119
Oboler, Arch *6:* 1066
Obstetrics *5:* 908; *6:* 1110
Oceanography *2:* 321; *4:* 705, **757-758**
Oden, Svante *1:* 10
Offshore drilling *4:* 763
Ohain, Hans von *1:* 34
Ohm, Georg *2:* 388
Oil *3:* 491, 561; *4:* 656, 694; *5:* 821
Oilcloth *6:* 1143
Oil drilling equipment *2:* 398
Oil engine *3:* 587
Oil lamp *3:* 491
Oil refining *3:* 493; *4:* **759-764**; *5:* 845
Oldham, Richard Dixon *2:* 369
Olds, Ransom Eli *4:* 680
Olduvai Gorge *3:* 559
Olestra *3:* 448
Olivetti *6:* 1104
olo *6:* 1036
Ololiuqui *3:* 522
On the Origin of Species by Means of Natural Selection *3:* 557; *5:* 812
Oort, Jan *5:* 946
Oort cloud *5:* 946
Open-angle glaucoma *3:* 508
Ophthalmia neonatorum *2:* 409
Ophthalmoscope *2:* 301
Opium *4:* 713
Optical disk *2:* 290, 305; *4:* 696, 699, **764-765**; *6:* 1130, 1132
Optical scanners *4:* 696
Orbiting Solar Observatories *6:* 1111
Orentriech, Norman *3:* 521
Organization of Petroleum Exporting Countries *5:* 823
The Origin of Continents and Oceans *2:* 319
O-rings *6:* 985
Orlon *1:* 13
Osborne, Thomas B. *6:* 1137
Oscillator *4:* 700, **765**; *5:* 866, 921
Oscilloscope *2:* 258
Osmium *4:* **765-766**; *5:* 845
Osmosis *4:* **766-767**
Otis, Elisha Graves *2:* 391; *5:* 932
Otto, Nikolaus *3:* 585
Outcault, Richard F. *2:* 284
Oval window *1:* 13
Ovum *3:* 453
Owens Illinois Glass Company *3:* 456
Ox eye *3:* 508
Oxygen *2:* 241, 282, 330; *3:* 450, 481, 514, 590, 591; *4:* 675, 694, 726, **767-770,** 771; *5:* 836, 858, 923, 924, 941, 949; *6:* 1033, 1090, 1124, 1128, 1169
Ozone *2:* 246; *3:* 482, 487, 514; *4:* **770-774**; *5:* 827; *6:* 1111, 1126, 1163
Ozone layer *1:* 20, 29; *3:* 470, 515

P

Pacemaker *5:* **805-806**
PacMan *6:* 1130
Paddle wheel *3:* 453
PageMaker *2:* 296
Painkillers *2:* 393
Pain relief *1:* 15; *3:* 571
Paint *5:* 925
Paleomagnetism *2:* 370
Palladium *5:* 845
Palmer, Henry *6:* 1085
Pampers *2:* 344
Panama Canal *2:* 233; *6:* 1172
Pancreas *5:* 914
Pancreozymin *5:* 916
Pangaea *2:* 319
Papanicolaou, George Nicholas *5:* 806
Paper *3:* 476; *4:* 701; *5:* 828, 856, 884, 926
Papin, Denis *2:* 237; *6:* 1017
Pap test *5:* **806-807**
Parachute *4:* 756
Parasite *2:* 246; *5:* 933
Parasol *6:* 1115
Paré, Ambroise *2:* 341; *3:* 478
Parker, David *6:* 1142
Parker, Eugene N. *5:* 948
Particle accelerator *5:* **807-810,** 873; *6:* 1024, 1036, 1167
Particle spin *4:* 751; *5:* **810**
Passive restraints *1:* 27
Pasteur, Louis *3:* 505, 574, 577; *4:* 707; *5:* 842; *6:* 1064, 1134, 1170
Patent medicine *2:* 278
Pathé, Charles *4:* 717
PATREC *2:* 317
Pattern recognition *2:* 317
Pauli, Wolfgang *5:* 866
Pauling, Linus *4:* 709; *5:* **810-812**
Pavlov, Ivan Petrovich *5:* **812-814,** 914
Pay-per-view broadcasting (Pay TV) *2:* 225
Pearce, Louise *5:* 934
Pearson, Charles *6:* 1029
Peganum harmala *3:* 522
Peking man *3:* 559
Pellagra *4:* 740; *6:* 1137
Pen *2:* 310; *5:* **814-815**; *6:* 1038, 1101
Pencil *2:* 400; *5:* 906
Pendulum *3:* 483
Penicillin *2:* 263; *3:* 511; *5:* **816-818,** 848; *6:* 1007, 1031, 1044, 1162, 1166
Pen plotters *2:* 310
Penzias, Arno *6:* 1013
Pepsin *2:* 347; *5:* 915
Percussion drilling *4:* 762
Peridotite *2:* 372
Periodic law *2:* 391; *3:* 594; *5:* **818-821**; *6:* 1048
Peripheral nervous system *4:* 731
Periscope *6:* 1026
Perkins, Jacob *3:* 534
Permanent waves *3:* 519
Perrier, C. *6:* 1048

Personal computer (PC) *2:* 405; *4:* 673
Peru *2:* 277
Peter, Daniel *2:* 266
Peterssen, S. *6:* 1149
Petrochemicals *3:* 562; *4:* 764
Petroleum *2:* 238; *3:* 509, 561; *4:* 723; *5:* **821-824**
Peyrere, Isaac de la *3:* 556
Pfaff, Philip *2:* 341; *3:* 444
Pfleumer, Fritz *4:* 668
Phagocyte *5:* **824-825**
Phenergan *6:* 1087
Phenotype *3:* 495
Phenylalanine *5:* 837
Phenylketonuria *2:* 399
Philbrook, B. F. *2:* 341
Philips Company *6:* 1132
Philosopher's stone *5:* 825
Phipps, James *3:* 577
Phocomelia *6:* 1062
Phoenicians *1:* 43
Phonofilm *4:* 715
Phosphorescence *5:* 826
Phosphorus *5:* **825-826,** 876, 887; *6:* 1089
Photochemistry *5:* **826-827**; *6:* 1114
Photochromoscope camera *2:* 279
Photocomposition *5:* 836
Photocopying *5:* **827-828**
Photoelectric cell *3:* 450; *6:* 1056
Photoelectric effect *2:* 375, 389; *4:* 655; *5:* **828-830,** 867
Photograph *3:* 449; *5:* 927; *6:* 1046
Photographic film *4:* 680; *5:* 834
Photography *2:* 372; *3:* 462, 549, 579; *4:* 640, 648, 716; *5:* 826, **830-834,** 842; *6:* 987, 1055, 1116
Photon *4:* 656; *5:* 830, 866, 867, 941
Photosphere *6:* 1033
Photosynthesis *2:* 242; *3:* 472; *4:* 771; *5:* 825, 826, **835-836**
Phototypesetting *5:* **836**
Photovoltaic cell *5:* 829
Physical globe *6:* 1045
Physics *2:* 282; *3:* 511, 526, 547
Pickling *3:* 473
Picornaviruses *6:* 1135
Pictures *2:* 278
Pierce, Ebenezer *3:* 525
Piezoelectric effect *5:* **836-837**
Pilots *1:* 31
Pinaud, E. D. *3:* 519
Pinna *1:* 12
Pioneer 10 *5:* 946, 949; *6:* 995
Pioneer 11 *5:* 946; *6:* 995
Pioneer Venus 1 *6:* 1129
Pioneer Venus 2 *6:* 1129
Pistol *4:* 679
Piston *3:* 584; *6:* 1017
Pitch *4:* 760; *5:* 895
Pitchblende *6:* 1117
Pituitary gland *1:* 14; *6:* 1094
Pixels *3:* 450
PKU (phenylketonuria) *2:* 399; *5:* **837-838**
Plague *4:* 737
Planck, Max *5:* 830, 865, 867
Planck's constant *5:* 830
Plane *1:* 29; *3:* 564
Planetary motion *4:* 738; *5:* **838-841**
Planets *3:* 512
Planet X *5:* 946
Plants *2:* 242, 401; *3:* 453
Plasma *4:* 750; *5:* **841-842**
Plastic *1:* 14; *2:* 237, 255, 303, 318, 335, 344, 400, 401, 412; *3:* 445, 456, 460, 476, 554; *4:* 633, 644, 701, 728, 764; *5:* **842-843,** 852-854, 884, 913, 928, 935, 942; *6:* 1015, 1037, 1050, 1100, 1115, 1143
Plastic deformation *2:* 379
Plate glass *6:* 1152
Platelets *3:* 539
Plate tectonics *2:* 321, 363, 370, 372; *4:* 705, 757, 758; *5:* **843-844**; *6:* 1129
Platinum *1:* 7; *3:* 444; *4:* 741, 766; *5:* **844-845**; *6:* 1159
Playfair, John *6:* 1117
Pleated fan *3:* 447
Plexiglas *1:* 14; *2:* 412
Plucknett, Thomas *4:* 643
Plumbism *4:* 645
Pluto *5:* **845-847,** 946
Pluto Platter *3:* 480
Plutonium *4:* 730
Pneumatic tire *5:* 906, 937
Pneumocystis carinii *1:* 22
Pneumonia *1:* 22; *5:* 818, **847-848**
Poincaré, Jules-Henri *6:* 1076
Point set topology *6:* 1075
Poison *2:* 243, 261
Polacolor *2:* 279
Polar fronts *6:* 1149
Polarized light *4:* 633
Polaroid Corporation *6:* 1066
Polaroid Land camera *3:* 579
Polio *1:* 20; *3:* 577; *4:* 767; *5:* **849-850**; *6:* 1135
Poliomyelitis. *See* Polio
Political globe *6:* 1045
Pollution *1:* 9; *2:* 243, 336; *3:* 513; *5:* 888
Polonium *5:* 881
Polychlorinated biphenyls (PCBs) *5:* **850-851**
Polyester *2:* 252, 335; *5:* **851-852,** 853; *6:* 1115
Polyethylene *2:* 335, 401; *3:* 554; *5:* 843, **852-853**; *6:* 1050, 1154
Polygas *1:* 20
Polygraph *4:* 649
Polymer *3:* 459, 470, 520; *4:* 634, 761; *5:* 842, 852, 854
Polymer and polymerization *2:* 336, 343, 345; *4:* 761; *5:* 852, **853-854,** 855, 913; *6:* 1100, 1133
Polypropylene *5:* 843, 853, **854**
Polystyrene *5:* 853, **854-855**
Polythene *5:* 852
Polyurethane *5:* 853, **855**
Poniatoff, Alexander M. *6:* 1131
Pop *5:* 938
Popsicle *3:* 480
Population genetics *2:* 404; *5:* **855-856**
Porcelain *2:* 341; *3:* 444
Porta, Giambattista della *5:* 831
Porter, Rufus *3:* 464

Boldfaced numbers indicate main entry pages; italicized numbers indicate volume number

Portier, Paul *3:* 574
Positron *2:* 390; *6:* 1022
Post-it note *5:* **856**
Poster paint *6:* 1104
Post, Wiley *1:* 33
Potassium *5:* **856-859**
Potassium inhibition *5:* 858
Pott, Percivall *2:* 248
Power plant *2:* 243
Power supply *2:* 287
Pox *6:* 1042
Pratt, John *6:* 1102
Precipitation *6:* 1149
Prenatal surgery *5:* **859-860**
Presbyopia *2:* 408, 410
Preservation *2:* 234
Preservative *1:* 4
Pressey, Sydney L. *6:* 1047
Pressure cooker *2:* 237
Priestley, Joseph *2:* 245, 400; *4:* 768; *5:* 906
Prime Meridian *6:* 1067
Principals of Geology *6:* 1117
Printers *2:* 310
Printing technology *1:* 28
Prion *6:* 1135
Prism *2:* 281; *4:* 652; *6:* 1006, 1054
Privy *6:* 1073
Probability *3:* 485
Procter & Gamble (P&G) *2:* 344; *3:* 448, 469
Programming languages *2:* 300
Projean, Xavier *6:* 1101
Project Mohole *2:* 372
Prominences *6:* 1033
Prontosil *6:* 1031
Propane *2:* 401
Propellant *1:* 19
Property *4:* 657
Protactinium *5:* **860-861**
Protein *1:* 7; *2:* 259, 353, 398; *3:* 447, 498, 539, 542, 560, 583; *4:* 656, 688, 709, 718, 730; *5:* 812, 890, 912, 916, 919; *6:* 1031, 1094, 1133
Proton *3:* 564; *4:* 734, 748, 749; *5:* **861-862,** 863; *6:* 1022, 1090
Protoplasm *4:* 741
Proust, Joseph Louis *6:* 1124
Prout, William *3:* 563
Psychrometry *1:* 28
Psylocybe mexicana *3:* 522
Ptolemy *5:* 943
Ptyalin *2:* 346
PUFF *2:* 405
Pulley *2:* 340; *3:* 479; *4:* 762; *6:* 1016
Pullman, George *6:* 1083
Pulsar *5:* **863,** 879; *6:* 1162
Pumice *2:* 400
Pump *2:* 249
Punnet, Reginald Crundall *2:* 404
Pupils *2:* 254
Purcell, Edward *4:* 752
Purkinje, J. E. *3:* 461
Pyrex *6:* 1054
Pyroxylin cements *1:* 17

Q

Qasim, Abu al *3:* 539
Quanta *5:* 867
Quantivalence *6:* 1124
Quantum *5:* 867
Quantum mechanics *3:* 526; *4:* 709; *5:* **865-866,** 868
Quantum numbers *5:* 810
Quantum theory *2:* 377, 389; *5:* **867-868**
Quark *5:* 861, 862; *6:* 1024
Quartz *4:* 651; *6:* 1114
Quartz-polaroid monochromatic interference filter *5:* 947
Quasar *5:* 863, **868,** 880; *6:* 1013, 1162
Quimby, Harriet *1:* 31
Quinine *5:* 816

R

Rabies *3:* 508, 584; *6:* 1134
Rabi, Isodor Isaac *4:* 752
Rabinow, Jacob *2:* 303
Raceway *6:* 1010
Radar *2:* 333; *3:* 503, 551; *4:* 702, 765; *5:* **869-872**; *6:* 1014, 1109, 1148
Radial keratotomy *2:* 407; *5:* **872**
Radiation *3:* 532; *4:* 736; *6:* 1111, 1126
Radiation detector *2:* 406; *5:* **872-873**
Radio *2:* 287; *3:* 450, 459; *4:* 646, 696, 697, 700-702, 765; *5:* 869, **874-876,** 900, 920, 921, 925, 933; *6:* 1046, 1049, 1056, 1141, 1148
Radioactive dating *4:* 692
Radioactive tracer *5:* **876-877,** 878, 881; *6:* 1048
Radioactivity *2:* 331, 389; *4:* 651; *5:* 827, **877-878,** 881, 882; *6:* 1090, 1118
Radio astronomy *5:* 863, 868, **879-880**
Radio Corporation of America (RCA) *3:* 450; *6:* 1058
Radio interferometer *5:* **880-881**
Radio navigation *4:* 725
Radio telescope *5:* 879
Radiotherapy *5:* **881**; *6:* 1162, 1167
Radio waves *2:* 388; *3:* 588; *4:* 671, 700, 752
Radium *5:* **881-882**
Radon *3:* 539; *4:* 726; *5:* **882-883**; *6:* 1159
Railroad *6:* 1096
Ramsay, William *3:* 538; *4:* 637, 726; *6:* 1159
Rand Corporation *2:* 348
Ranger 7 *6:* 995
Ranger 8 *6:* 995
Ranger 9 *6:* 995
Rask, Grete *1:* 22
Rayleigh, Lord *4:* 637, 726; *6:* 1145
Rayon *2:* 335, 344; *3:* 459; *5:* 842, 853
Real number *3:* 593; *5:* **883-884**
Reaper and binder *4:* 680
Reber, Grote *5:* 879
Receiver *3:* 528
Receptors *4:* 731
Recessive gene *3:* 494
Rechendorfer, Joseph *2:* 400
Record *6:* 1133
Recycling *5:* **884-886**
Red giant *4:* 746; *6:* 1033
Red-shift *5:* **886-887**
Red tide *5:* **887-888**
Reed, Walter *6:* 1172
Reflection hologram *3:* 551
Reflectors *6:* 1054
Refrigeration *1:* 27; *2:* 330; *3:* 475, 572

Reitz, Bruce *6:* 1094
Relativity *3:* 512, 526; *5:* **888-890**
Remington Fire Arms *6:* 1103
Remington Rand Company *2:* 310
Repeaters *6:* 1050
Reproduction *2:* 258
Reserpine *3:* 570; *6:* 1087
Resin *1:* 17
Respiration *2:* 241; *5:* 836
Restriction enzyme *2:* 399; *5:* **890**
Retinal scanner *5:* 891
Retinitis *2:* 410
Retinography *5:* **891**
Retrieval *2:* 308
Retrovirus *1:* 24
Reverdin, Jacques Louis *6:* 1091
Reverse osmosis *2:* 343
Revolvers *3:* 518
Rhazes *3:* 575
Rh factor *5:* 860; **891-892**
Rhodium *5:* 845
RhoGam *5:* 892
Richardson, Ithiel *3:* 463
Richet, Charles *3:* 574
Richter, Charles F. *2:* 366
Richter scale *2:* 366
Rickets *6:* 1114
Rickover, Hyman *6:* 1028
Rifle *3:* 517
Ritter, Johann *6:* 1112
Ritty, James J. *2:* 253
Ritty's Incorruptible Cash Register *2:* 253
RNA (ribonucleic acid) *2:* 259, 356; *5:* **893**
Road building *5:* **893-897**
Robbery *2:* 252; *4:* 657
Robbins, Benjamin *6:* 1155
Robotics *2:* 307, 333; *4:* 697; *5:* **897-900**
Rochow, E. G. *5:* 924
Rocket *2:* 292, 406; *4:* 771; *6:* 997
Rocket and missile *1:* 36; *2:* 333; *5:* **900-904,** 941; *6:* 1078, 1149
Rollerskate *5:* 929
Roman candle *3:* 465
Roman numerals *5:* **904-905**; *6:* 1176
Röntgen, Wilhelm *6:* 1160, 1163, 1165, 1167
Roosevelt, Franklin D. *2:* 376
Root, Elisha King *4:* 679
Rope *4:* 633, 756
ROSAT *6:* 1005
Rossby, Carl-Gustaf *4:* 692
Rossi, Bruno B. *5:* 948
Rotary drilling *4:* 762
Rotary excavator *6:* 1097
Rotational molding *5:* 854
Rotheim, Eric *1:* 19
Rouelle, Jean *6:* 1120
Rubber *1:* 17; *2:* 400; *5:* 855, 937; *6:* 1143, 1177
Rubber bands *5:* 905
Rubber cement *1:* 17
Rubber, vulcanized *3:* 444; *5:* **905-906**
Rubidium *6:* 1007
Rubik, Erno *5:* 907
Rubik's cube *5:* **907**
RU 486 *5:* **907-908**
Rugby *3:* 476
Rugs *2:* 250
Rum *1:* 37
Russell, Elizabeth Shull *4:* 718
Russell, Sidney *2:* 381
Ruthenium *5:* 845
Rutherford, Ernest *4:* 734; *5:* 862; *6:* 1022, 1090
Ryle, Martin *5:* 880

S

Sabin, Albert *5:* 850
Sabin, Florence Rena *4:* 662
Saccharomyces cerevisiae *3:* 451
Sachs, Julius von *3:* 569
Safe *5:* **909-910**
Safer *6:* 993
Safety *1:* 26; *2:* 329
Safety match *5:* 826
Sagan, Carl *5:* **910-912**
Sales receipt *2:* 252
Salk, Jonas *1:* 21; *3:* 577; *5:* 850
Sal mirabile *3:* 563
Salmon, Daniel *3:* 574
Salt *2:* 342; *5:* 939
Saltpeter *5:* 856, 939
Salvarsan *6:* 1043
Salyut *6:* 987
Salyut 1 *6:* 1001
Salyut 3 *6:* 1001
Samarium *5:* **912**
Sampling *6:* 1041
San Andreas fault *5:* 918
Sandpapers *1:* 1
San Francisco earthquake *5:* 917
Santa Anna, Antonio López de *2:* 265
Santos-Dumont, Alberto *1:* 31
Saran *5:* **913**
Sargent, James *5:* 909
Satellite *2:* 287, 406; *3:* 484, 487; *4:* 693; *5:* 876; *6:* 1005, 1050
Sater, Mel *6:* 1131
Saturated fats *3:* 447
Saturn *3:* 484, 566; *4:* 694, 729, 738; *5:* 846, 912, **913-914,** 946; *6:* 995, 1118
Savery, Thomas *6:* 1017
Scaling *2:* 343
Scanner *3:* 449
Schally, Andrew Victor *6:* 1094
Schaudinn, Fritz *6:* 1043
Scheele, Carl Wilhelm *2:* 283; *3:* 509; *6:* 1095
Scheelite *6:* 1095
Schertel, Hans von *3:* 564
Schiaparelli, Giovanni *4:* 691
Schlanger, Jay *2:* 409
Schmidt, Bernhard *6:* 1055
Schmidt, Maarten *6:* 1013
Schönbein, Christian Friedrich *5:* 842
Schrieffer, J. Robert *6:* 1035
Schrödinger, Erwin *5:* 865; *6:* 1099
Schultz, Charles *2:* 284
Schutz, H. H. *1:* 28
Schwann, Theodor *6:* 1070
Schweitzer, Hoyle *6:* 1154
Scientific theory *3:* 482
Scissors *2:* 340
Scooter *5:* 929
Scotch-gard *6:* 1144
Scotch tape *1:* 18
Screw *3:* 479, 537; *6:* 1016
Screwdriver *6:* 1038
Screw propeller *6:* 1024

Boldfaced numbers indicate main entry pages; italicized numbers indicate volume number

Scribner, Belding *2:* 344
Scrubbers *1:* 11
Scuba diving *2:* 350
Scurvy *6:* 1138
Sea floor spreading *4:* 757; *5:* 843
Seaplane *1:* 31
Seasat I *2:* 374
Seat belts *1:* 26
Secchi, Pietro Angelo *6:* 1007
Secondary glaucoma *3:* 508
Second law of thermodynamics *3:* 531
Secretin *3:* 553; *5:* **914-916**
Secretinase *5:* 916
Sega Genesis *6:* 1130
Segré, Emilio *6:* 1023, 1048
Seismograph *2:* 366, 372; *4:* 707; *5:* 917
Seismology *2:* 365; 368, 371; *5:* **916-918**
Seismometer *5:* 917
Selenium *5:* **918-919**; *6:* 1064
Seltzer *2:* 241
Semiconductor *5:* 919; *6:* 1099
Serum hepatitis *3:* 542
Set theory *6:* 1076
Sewage systems *6:* 1073
Sex chromosome *2:* 271; *3:* 494; *5:* **919-920**
Sex hormones *2:* 264; *4:* 657
Sextant *4:* 725
Sexually transmitted disease *6:* 1041
Sexual reproduction *2:* 276
Shakespeare, William *6:* 1119
Shampoo *3:* 520
Shannon, Robert V. *3:* 528
Shearer *4:* 643
Sheffield, Washington W. *6:* 1075
Shelter *3:* 502
Shepard, Alan *6:* 981
Sherman, J. Q. *1:* 28
Sherman, Patsy *6:* 1144
Sherrington, Charles Scott *6:* 1040
Shock waves *2:* 369, 372; *4:* 706; *5:* 917
Shoemaker-Levy 9 *3:* 599, 601; *6:* 1006
Sholes, Christopher Latham *6:* 1102
Shooting star *4:* 689
Short radio waves *2:* 280
Shortwave radio *5:* **920-921**
shoyu *3:* 451
Shumway, Norman *6:* 1092, 1094
Sickle-cell anemia *3:* 500
Sickness *2:* 339
Sidera Medicea *3:* 484
sifr *6:* 1175
Sight *2:* 318, 410
Signature *3:* 460
Sign language *5:* **921-923**
Sikorsky, Igor *1:* 33; *3:* 536
Silent Spring *2:* 338
Silica *3:* 457
Silicon *4:* 647; *5:* 842, 900, **923-924**
Silicone *5:* 853, **924-926**; *6:* 1143
Silicone rubber *5:* 925
Silk *3:* 443; *4:* 755
Silk-screen printing *5:* **926-927**
Silly Putty *5:* 927-928
Silver *3:* 444; *4:* 684, 741, 752; *5:* **928-929,** 941; *6:* 1088
Simon, Samuel *5:* 927
Simons, E. *5:* 854
Simple Pleasures *3:* 448
Simplesse *3:* 448
Singer, Norman *3:* 448
Singularity *3:* 527
Sinusoidal *4:* 765
Skateboard *5:* **929**
Skegs *6:* 1037
Sketchpad *2:* 305
Skin diving *2:* 350
Skin grafting *6:* 1091
Skis and ski bindings *5:* **930-931**
Skylab *5:* 947; *6:* 1001, 1003, 1111
Skyscraper *2:* 391; *5:* **931-932**
Skywriting *5:* **932-933**
Sleeping *2:* 381
Sleeping sickness *5:* **933-934**
Slingshots *3:* 480
Slinky *5:* **934-935**
Slot machine and vending machine *5:* **935-937**
Small Astronomy Satellite (SAS) *6:* 1005
Small intestine *2:* 347
Smallpox *3:* 573, 575; *6:* 1135
Smelting *5:* 928
Smith, Hugh *6:* 1173
Smith, Oberlin *4:* 668
Smith, Robert Angus *1:* 9
Smith, Samuel *6:* 1144
Smith, Theobald *3:* 574
Smith, William *4:* 762
Snorkel *6:* 1028
Snowmobile *5:* **937-938**
Soap *3:* 520
Soap and detergent *6:* 1142
Sobrero, Ascanio *2:* 360
Soccer *3:* 476
Soda *5:* 941
Soda fountain *5:* 938
Soda pop *5:* **938**
Soda water *2:* 241; *5:* **938**
Sodium *3:* 563; *4:* 742; *5:* 857, **939-941**
Sodium salvarsan *6:* 1044
Soft contact lenses *2:* 319
Soft drinks *5:* 938
Software *2:* 295
Solar collector *3:* 535
Solar distillation *2:* 343
Solar-flare activity *2:* 371
Solar flares *6:* 1033
Solar-powered machines *5:* 829
Solar sail *5:* **941-942**
Solar system *3:* 484, 488, 591; *4:* 687, 694, 728, 730, 757; *5:* 868, 913, **943-946,** 949; *6:* 1118
Solar telescope *5:* **946-947**; *6:* 1055
Solar wind *2:* 370; *5:* **947-949**; *6:* 1033, 1126
Solder *6:* 1068
Soluble CD4 *1:* 25
Solutions *1:* 8
Somatotropin *3:* 553
Sonar *5:* **949-950**; *6:* 1152
Sonar depth finders *4:* 704
Sony Corporation *6:* 1132, 1141
Soulé, Samuel W. *6:* 1102
Sound *1:* 12; *2:* 385
Sound barrier *1:* 35
Sound recording *4:* 640, 668
Sound waves *1:* 12
Soviet Union *2:* 406; *6:* 1067
Soy sauce *3:* 451
Soyuz *6:* 986
Soyuz 1 *6:* 986
Soyuz 11 *6:* 987

Soyuz 24 *6:* 1001
Space *2:* 372
Spaceball *6:* 1086
Spacecraft *2:* 294; *5:* 941
Spacecraft, manned *6:* **981-988**
Space equipment *6:* **988-993**
Space exploration *2:* 406
Space Infrared Telescope Facility (SIRTF) *6:* 1005
Space probe *4:* 704; *5:* 900, 904; *6:* **994-997,** 1055
Space race *2:* 406
Space shuttle *1:* 36; *2:* 373; *6:* 985, 991, **997-1000,** 1003, 1005
Space station *6:* 987, **1000-1003**
Space telescope *6:* **1003-1006,** 1055
Spark chamber *5:* 873
Spark plugs *5:* 925
Spear *3:* 524
Spectacle *6:* 1052
Spectral lines *5:* 810
Spectroheliograph *5:* 946
Spectrohelioscope *5:* 946
Spectrometer *6:* 1006
Spectroscope *3:* 551; *5:* 946; *6:* 1064
Spectroscopic lines *5:* 866
Spectroscopy *2:* 282; *4:* 656, 726, 752; *6:* **1006-1007,** 1032, 1065, 1128
Spectrum *6:* 1006
Specular reflection *4:* 652
Speech *2:* 348
Speech recognition *2:* 315
Speed *2:* 329
Speedostat *2:* 329
Sperm *3:* 453
Sperry, Elmer *2:* 292
Sperry, Roger W. *6:* 1008
Sperti, George Speri *2:* 318
Sphere of Aeolus *5:* 900
Sphygmomanometer *4:* 649
Spinal tap *4:* 683
Spinning wheel *2:* 341
The Spirit of St. Louis *1:* 33
Spirits *1:* 37
Spleen *4:* 662
Splint *3:* 478
Split-brain functioning *6:* **1008-1009**
Spontaneous combustion *2:* 284
Spontaneous emission *4:* 641
Sports *3:* 476; *5:* 929, 930; *6:* 1036
Spring *2:* 293; *3:* 537; *5:* 935; *6:* 1038, 1085
Springweight scales *2:* 293
Sputnik *5:* 904
Spy satellite *6:* 1055
S. S. White Company *6:* 1075
Stage decompression *2:* 339
Stahl, Georg Ernst *1:* 3
Standard Model *5:* 809
Standard Oil *3:* 491
Stannous fluoride *3:* 470
Stapes *1:* 13
Stapler *6:* **1009-1010**
Star cluster *6:* **1010-1012**
Starling, Ernest *5:* 914
Stars *3:* 487
Starzl, Thomas *6:* 1093
Static electricity *2:* 387
Steady-state theory *6:* **1012-1013**
Stealth aircraft *1:* 36; *6:* **1013-1016**
Steamboat *6:* 1018, 1026
Steam engine *1:* 29; *2:* 332, 349, 396, 397; *3:* 453, 584; *4:* 762; *6:* **1016-1018,** 1026, 1080, 1156
Steel *1:* 41; *2:* 291; *4:* 762, 768; *6:* 1115
Steele, Thomas *2:* 342
Steinmetz, Charles *1:* 46
Stellar evolution *3:* 487, 527
Stem cells *5:* 860
Stephenson, George *6:* 1080
Stereo *5:* 900; *6:* **1019**
Stereoisomer *3:* 593
Sterilizing *2:* 348
Sterling silver *5:* 928
Steroid *2:* 269, 322
Stetigkeit und irrationale Zahlen *5:* 884
Steven, A. Edwin *3:* 528
Stevens, John *6:* 1018
Stevens, Nettie *5:* 920
Stevens, Robert *6:* 1081
Still *1:* 37
Stimulant *3:* 580
Stirdivant, A. H. *2:* 352
Stockings *4:* 755
Stockton, Samuel *3:* 444
Stoddard, W. J. *2:* 358
Stomach *2:* 347; *5:* 914
Stomata *5:* 835
Storage *2:* 302, 308
Stories *2:* 284
Storm *6:* **1020-1022,** 1147
Stourbridge Lion *6:* 1080
Stove *3:* 519, 533
Strabismus *2:* 408
Stranger in a Strange Land *6:* 1143
Strasburger, Eduard *2:* 260
Stress *1:* 14
Structural metals *4:* 667
Structure *3:* 502
Strutt, John William *4:* 637, 726; *6:* 1145
Subatomic particle *2:* 382; *3:* 527; *5:* 862; *6:* **1022-1024,** 1090
Submarine *2:* 292; *5:* 949; *6:* **1024-1028,** 1078, 1110
Subroc *6:* 1078
Subtraction *1:* 38; *2:* 227, 228
Subway *6:* **1028-1030**
Suction *6:* 1017
Suez Canal *2:* 233
Suffocation *2:* 243
Sugarfree gum *2:* 266
Sulfa *5:* 816
Sulfonamide drug *2:* 263; *4:* 683; *6:* **1030-1031**
Sulfur *2:* 400; *3:* 565; *4:* 675, 686, 725; *5:* 906, 940; *6:* 1125
Sulfur dioxide *1:* 10
Sulfuric acid *1:* 7; *3:* 563, 565; *5:* 825, 845
Sun *2:* 240, 280; *3:* 487, 537, 590, 591; *4:* 653, 668, 686, 692, 773; *5:* 838, 845, 943, 944, 947; *6:* 1006, **1031-1034,** 1070, 1111, 1113, 1118, 1126, 1127, 1144
Sundback, Gideon *6:* 1176
Sundial *2:* 272, 293
Sunspot *3:* 484; *5:* 948; *6:* 1033, 1126
Sunya *6:* 1175
Supercomputer *2:* 298; *6:* **1034-1035**

Boldfaced numbers indicate main entry pages; italicized numbers indicate volume number

Superconducting quantum interference devices (SQUIDS) *6:* 1100
Superconducting Super Collider (SSC) *5:* 809; *6:* 1036
Superconductivity *2:* 331; *3:* 538; *5:* 809, 866; *6:* **1035-1036,** 1099
Superglue *1:* 19
Super magnets *2:* 292
Supernova *3:* 488; *5:* 863
Supernova 1987A *4:* 747
Superphosphate *5:* 825
Supramar *3:* 564
Surfboard *6:* **1036-1037,** 1154
Surgery *2:* 254, 394; *3:* 529; *4:* 641
Surgical transplant *2:* 384
Surveyor *6:* 994
Sutcliffe, R. C. *6:* 1149
Sutter, Eugene *3:* 519
Sutton, Walter S. *2:* 271, 404; *3:* 546
Swammerdam, Jan *4:* 662
Swan, Joseph *3:* 459
Sweets *2:* 266
Swiss army knife *6:* **1038**
SX-70 color film *3:* 580
Sylvinite *5:* 858
Sylvius, Franciscus *1:* 7
Synapse *1:* 5; *4:* 734; *6:* **1038-1040**
Syncom *2:* 288
Synthesizer, music *6:* **1040-1041**
Synthesizers *4:* 719
Synthetic *1:* 13; *2:* 335
Synthetic fiber *4:* 764
Synthetic rubber *5:* 927
Syphilis *2:* 263; *5:* 818; *6:* **1041-1044**
Systema naturae *3:* 557
System of the Earth. *6:* 1116
Szilard, Leo *4:* 748, 749, 753

T

Tackers *6:* 1010
TACVs *6:* 1084
Taggart, William H. *2:* 341
Tagliacozzi, Gaspare *6:* 1090
Takemine, Jokichi *3:* 552
Talbot, William Henry Fox *3:* 549; *5:* 833
Talbotype *5:* 833
Tallow *3:* 447
Tape *4:* 668
Tape dispenser *1:* 18
Tape recorder *4:* 668
Tape recordings *6:* 1046
Tassel, James Van *2:* 230
Taveau, Auguste *2:* 341
TAXMAN *2:* 405
Taxol *2:* 264
Tay-Sachs disease *3:* 500
Teaching aid *6:* **1045-1047,** 1055
Teaching machines *6:* 1047
Technetium *6:* **1048**
Technicolor Corporation *4:* 715
Technology *2:* 276, 292, 294, 315, 317; *3:* 535
Teeter, Ralph *2:* 329
Teeth *2:* 340, 341
Teflon *2:* 344; *3:* 470; *6:* 991, 1144
Telecommunications *3:* 458
Telegraph *3:* 450; *6:* 1051, 1066, 1147, 1148
Telegraph cable *5:* 853
Telephone *2:* 286, 308, 311, 316; *3:* 449, 458, 528; *4:* 695, 698, 699; *6:* 1019, 1048, 1049, 1148
Telephone answering device *6:* **1048-1049**
Telephone cable *2:* 286
Telephone cable, transatlantic *6:* **1049-1051**
Telephone-Transmitter *3:* 528
Teleprinter and teletype *6:* **1051-1052**
Telescope *4:* 710; *5:* 880; *6:* 1003, 1056, 1119, **1052-1055**
Teletype *6:* 1052
Teletypewriter *6:* 1051
Television *2:* 258, 288; *3:* 459, 551; *4:* 700, 765; *5:* 829, 900; *6:* 985, 1046, **1056-1060,** 1129, 1131
Television and Infra Red Observation Satellite (TIROS 1) *6:* 1150
Telex *6:* 1052
Telex machines *6:* 1051
Teller, Edward *3:* 567
Telstar I *2:* 288
Temin, Howard *5:* 893
Temperature *1:* 2, 28
Temperature scale *1:* 3
Templeton, James *2:* 250
Tenements *6:* 1073
Tenite *3:* 480
Tennant, Smithson *4:* 766
Tension *2:* 379
Tequila *1:* 38
Teratogen *6:* **1060-1063**
Tereshkova, Valentina *6:* 982
Terry, Eli *4:* 680
Terylene *2:* 335
Tesla, Nikola *1:* 45; *5:* 876
Tetanus *6:* **1063-1064**
Textile *1:* 27; *2:* 335; *5:* 851, 926
Thales *2:* 386
Thalidomide *6:* 1062
Thallium *6:* **1064-1065**
Theft *2:* 252; *4:* 659
Theiler, Max *6:* 1173
Thenard, Louis *5:* 858, 940
Theory of Games and Economic Behavior *3:* 486
Theory of resonance *5:* 811
Theory of sea floor spreading *2:* 321
Theory of the Earth *3:* 481
Therapy *2:* 322, 393; *3:* 499
Theremin *4:* 719
Thermal (heated) hair rollers *3:* 520
Thermionic emission *2:* 389
Thermofax *5:* 828
Thermoforming *5:* 854
Thermography *5:* 827
Thermometer *3:* 530; *4:* 635, 644, 685
Thermonuclear reactions *3:* 567
Thermoplastic *5:* 853, 855, 913
Thermos *6:* 1123
Thermosets *5:* 853
Thermostat *2:* 332; *4:* 699; *6:* 1071
Third law of thermodynamics *3:* 532
Thomas, W. F. *2:* 284
Thomin *2:* 411
Thompson, William *6:* 1115
Thom, René *6:* 1077
Thomson, Joseph J. *2:* 382; *5:* 829, 862
Thomson, William *1:* 2; *2:* 388

Thorium *5:* 883
3-D *3:* 549
3-D motion picture *4:* 715; *6:* **1065-1066**
Thunderstorm *6:* 1020
Thurber, Charles *6:* 1101
Thyroxine *3:* 552; *6:* 1094
Tides *4:* 758
Time *2:* 271
Time-space continuum *2:* 375
Time zone *2:* 273; *6:* **1066-1067**
Tin *1:* 40; *3:* 470, 565; *4:* 645, 684; *6:* **1068,** 1088
Tin foil *2:* 341
Tire, radial *4:* 633
Tissue *1:* 24; *2:* 263; *3:* 447, 451; *4:* 731, 743, 752; *5:* 914; *6:* **1069-1070**
Titania *6:* 1119
Titanic *5:* 949
Titanium *4:* 712; *6:* **1070-1071**
TNT *4:* 741
Toaster *6:* **1071**
Toilet *6:* **1072-1073**
Toll House cookies *2:* 267
Tombaugh, Clyde *5:* 846
Tom Thumb *6:* 1081
Toothbrush and toothpaste *2:* 326; *3:* 460, 469; *6:* **1074-1075**
Tooth decay *3:* 468
Tooth extraction *3:* 443
Toothpaste *3:* 469
Topology *6:* **1075-1077**
Tornadoes *6:* 1021
Torpedo *6:* 1026, **1077-1078**
Torque *3:* 535
Torricelli, Evangelista *4:* 684; *6:* 1148
Toshiba *6:* 1131
Total internal reflection *4:* 652
Toys *3:* 467, 479, 554; *4:* 633, 635; *5:* 907, 927, 929, 934; *6:* 1085
Tracers *3:* 594
Trachoma *2:* 409
Tracked Air-Cushion Vehicles *6:* 1084
Traction *3:* 479
Tractor *2:* 398
Traffic accidents *1:* 26
Traffic signal *6:* **1079**
Train and railroad *3:* 453; *5:* 895; *6:* 1028, 1066, **1080-1085,** 1099
Training *2:* 312
Trait *2:* 271, 353; *3:* 493
Trampoline *6:* **1085-1086,** 1127
Tranquilizer (antipsychotic type) *3:* 570; *6:* **1086-1088**
Transducer *6:* 1107
Transformer *1:* 45
Transfusion *1:* 23
Transistor *2:* 300; *3:* 528; *4:* 647, 672, 697; *5:* 805
Transit *6:* 1128
Transmission hologram *3:* 551
Transmutation of elements *6:* **1088-1090**
Transplant, surgical *6:* **1090-1094**
Transportation *2:* 231, 250, 391, 394, 396; *3:* 491, 563
Transuranium *6:* 1090
Transverse waves *4:* 650
Travers, Morris *4:* 637; *6:* 1159
Trefouel, Jacques *6:* 1031
Trefouel, Thérèse *6:* 1031
Tremors *2:* 371
Trendar *3:* 572
Treponema pallidum *6:* 1043
Trevithick, Richard *6:* 1080
TRH (thyrotropin-releasing hormone) *6:* **1094**
Troposphere *4:* 692
Truck *2:* 398
Trudeau, Garry *2:* 286
Truth *4:* 649
Trypanosomiasis *5:* 933
Tryparsamide *5:* 934
Tryptophan *4:* 740
Tsetse fly *5:* 933
Tsiolkovsky, Konstantin *5:* 902, 941
Tsunamis *2:* 366
Tuberculosis *1:* 20; *3:* 505; *4:* 682
Tumors *6:* 1094
Tungsten *6:* **1095,** 1153
Tungstic acid *6:* 1095
Tunnel *3:* 451; *6:* 1030, **1095-1099**
Tunneling *6:* **1099-1100**
Tuohy *2:* 412
Tupper, Earl *6:* 1100
Tupperware *6:* **1100-1101**
Turboprop *3:* 595
Turing, Alan *1:* 38; *2:* 228
Turpentine *5:* 905
Turtle *6:* 1024
Twain, Mark *6:* 1103
Twins *2:* 275
Tympanic membrane *1:* 12
Typewriter *6:* 1051, **1101-1104**
Typhoon *6:* 1021
Typing correction fluid *6:* **1104-1105**

U

UHF *2:* 385
Uhlenbeck, George *5:* 810
Ultrasonic emulsification *6:* 1108
Ultrasonics *6:* 1145
Ultrasonic wave *6:* **1107-1108**
Ultrasonography *6:* 1109
Ultrasound device *6:* **1108-1110**
Ultrasound imaging *5:* 860
Ultraviolet *6:* 1111
Ultraviolet astronomy *6:* **1111-1112**
Ultraviolet radiation *2:* 282; *3:* 468, 515; *4:* 771, 773; *5:* 827; *6:* 1003, **1112-1114,** 1163
Ultraviolet radiation lamps *6:* 1112, 1160
Umbrella *6:* **1114-1115**
Umbria *6:* 1119
Uncertainty principle *5:* 865
Unconditioned reflex *5:* 813
Underwater photography *6:* **1115-1116**
Uniformitarianism *6:* **1116-1117**
UNIVAC *2:* 303; *4:* 666
Universe *3:* 526
University of California at Berkeley *5:* 917
University of Zürich *2:* 375
Unprotected sex *1:* 23
Upatnieks, Juris *3:* 551
Updrafts *6:* 1021

Boldfaced numbers indicate main entry pages; italicized numbers indicate volume number

Upjohn Company *3:* 571
Uranium *3:* 470; *4:* 676; *5:* 883; *6:* **1117-1118**
Uranus *4:* 695, 728, 730; *5:* 846, 946; *6:* 1117, **1118-1120**
Urban VIII *3:* 484
Urea *6:* **1120-1121**
Urey, Harold Clayton *3:* 565
U.S. Environmental Protection Agency *2:* 338
U.S. Explorer 48 *3:* 488
U.S. Food and Drugs Act *3:* 476
U.S.S. *Holland* *6:* 1026
U.S.S. *Housatonic* *6:* 1078
U.S. Weather Service *6:* 1148
Uterine cancer *5:* 807

V

Vaccine *1:* 25; *2:* 264, 269; *3:* 573; *4:* 717
Vacuum *6:* 1095, 1145
Vacuum bottle *6:* **1123-1124**
Vacuum flask *2:* 330
Vacuum tube *2:* 293, 300; *4:* 646, 672
Vagusstoffe *1:* 6
Valence *6:* **1124-1125**
Valve *3:* 530; *4:* 763
Van Allen, James *6:* 1125
Van Allen belts *6:* **1125-1126**
Van Allen radiation belts *2:* 406, 407
Van de Graaf, Robert *2:* 388
Van der Waals, Johannes *4:* 634
Vanguard II *6:* 1149
Van Helmont, Jan *2:* 241
Variola *3:* 575
Variolation *3:* 573, 575
Vegetables *3:* 569
Vehar, Gordon *3:* 498
Velcro *6:* **1126-1127**
Velocity *2:* 385
Vending machine *5:* 936
Venera 1 *6:* 1128
Venera 7 *6:* 995
Venera 9 *6:* 995, 1128
Venera 10 *6:* 995, 1128
Venturi, Robert *5:* 932
Venus *2:* 242; *3:* 484, 597; *5:* 910, 945; *6:* 995, 1118, **1127-1129**
Verrier, Urbain Le *5:* 845
Very Long Baseline Interferometry (VLBI) *5:* 881
VHF *2:* 385
Vibrio cholerae *2:* 268
Video game *4:* 697; *6:* **1129-1130,** 1139
Video recording *4:* 697; *6:* **1130-1132**
Videotape *4:* 670; *6:* 1046
Viking 1 *4:* 677; *6:* 995
Viking 2 *4:* 677; *6:* 995
Vinblastine *2:* 264
Vincristine *2:* 264
Vinyl *1:* 3; *2:* 381; *3:* 460; *5:* 854, 913; *6:* **1132-1133**
Viral interference phenomenon *3:* 583
Virtual particles *3:* 527
Virtual reality *2:* 313
Virus *1:* 21, 24; *2:* 246, 256, 263; *3:* 504, 541, 542, 575; *4:* 682; *5:* 848, 849, 890; *6:* **1133-1135,** 1171
Vision *2:* 318, 410
Vitamin *3:* 447; *4:* 657, 740; *5:* 812, 827; *6:* 1007, **1136-1138**
Vitamin B *3:* 471
Vitamin B family *3:* 471, 739
Vitamin C *3:* 476
Vitamin C and the Common Cold *5:* 812
Vitamin D *6:* 1114
Vitamin K *6:* 1137
VLF *2:* 385
Voder *6:* 1138
Vodka *1:* 38
Voice mail *6:* 1049
Voice Operation Demonstrator *6:* 1138
Voice synthesizer *6:* 1130, **1138-1139**
Volcanism *3:* 481
Volta, Alessandro *1:* 45; *2:* 380, 387; *5:* 939, 940
Voltaic pile *2:* 387
Voronoy, Yuri *6:* 1092
Voskhod *6:* 986
Vostok *6:* 985
Vostok 1 *6:* 986
Voyager *5:* 949; *6:* 1120
Voyager 1 *5:* 946
Voyager 2 *4:* 729; *5:* 946
VPL Research, *2:* 313
Vries, Hugo de *3:* 545; *4:* 721
Vucetich, Juan *3:* 461
Vulcanization *5:* 906

W

Waddell, W. E. *6:* 1066
Wakefield, Ruth *2:* 267
Waldeyer-Hartz, Wilhelm von *2:* 271; *4:* 734; *6:* 1039
Walker, Sarah Breedlove *3:* 520
Walkie-talkie *5:* 876
Walkman *4:* 669; *6:* **1141**
Wallace, Alfred Russell *2:* 403
Wang Laboratories *2:* 295
Wankel engine *1:* 31; *3:* 587
Warfare *2:* 260, 359, 377; *3:* 462, 465, 466, 470, 472, 485, 488, 516, 517, 535, 564, 567, 594; *4:* 736, 748; *5:* 826, 869, 900, 916, 927, 930, 949; *6:* 1013, 1019, 1024, 1077, 1117
Warmth *2:* 381
Warship *1:* 32; *6:* 1026
Washansky, Louis *6:* 1092
Washing *2:* 276, 348
Washing machine *2:* 277, 359; *4:* 642, 699; *6:* **1141-1142**
Washington, George *3:* 443
Wassermann, August von *6:* 1043
Water *2:* 342; *4:* 768
Waterbed *6:* **1143**
Water clock *2:* 272
Water closet *6:* 1073
Waterman, Lewis Edson *5:* 814
Waterproof material *5:* 905; *6:* 1133, **1143-1144**
Watson, James *2:* 354, 404; *3:* 547
Watson, Thomas J. *2:* 302; *3:* 530
Watson-Watt, Robert *5:* 869
Watt, James *2:* 332; *3:* 534; *6:* 1017, 1080
Waveform *6:* 1040
Waveguide *4:* 701, 703
Wavelength *2:* 385
Wave motion, law of *2:* 282; *4:* 651, 656; *6:* 1108, **1144-1145**

Wave theory *4:* 654
Wax *6:* 1143
Weapon *3:* 466, 488, 516, 517, 564, 567; *5:* 900
Weather *1:* 28
Weather *6:* 1149
Weather balloon *6:* 1148
Weather forecasting *4:* 694
Weather forecasting model *6:* 1022, 1055, **1145-1149**
Weather map *6:* 1148
Weather satellite *4:* 704; *6:* 1148, **1149-1151**
Weaver, Jefferson Hane *5:* 889
Wedgwood, Thomas *5:* 831
Wegener, Alfred Lothar *2:* 319
Weinberg, Wilhelm *5:* 855
Weinstein, James *6:* 1143
Weiss, Paul *6:* 1008
Wells, Horace *3:* 445
Wenham, Frank *6:* 1156
Werner, Alfred *4:* 707
Western Electric Company *4:* 715
Westinghouse, George *1:* 46; *6:* 1083
Westinghouse Electric Corporation *6:* 1058
Wham-O *3:* 480, 554
Wheel *3:* 451
Wheelchair *6:* **1151-1152**
Wheeler, Dr. Schuyler Skaats *3:* 446
Wheels and axles *6:* 1016
Whinfield, John *2:* 335; *5:* 852
Whipple, George *3:* 591
Whiskey *1:* 37
White dwarf star *6:* 1033
White, Ed *6:* 983
Whitehead, Robert *6:* 1078
White light *2:* 280; *6:* 1006
White out *6:* 1104
White, Tim D. *3:* 560
Whitney, Eli *4:* 679
Whittle, Frank *1:* 34; *3:* 595
Wichterle *2:* 412
Wiener, Norbert *2:* 333
Wig *3:* 520
Wildt, Rupert *6:* 1128
Wilkes, Maurice *4:* 666
Williams, John *6:* 1104
Williamson, Alexander W. *6:* 1125
Wilson, Edmund *5:* 920
Wilson, James *6:* 1046
Wilson, Robert W. *6:* 1013
Windshield *1:* 27; *5:* 937
Windshield and windshield wiper *6:* **1152-1154**
Windsurfer *6:* **1154-1155**
Wind tunnel *1:* 36; *6:* **1155-1157**
Wine *1:* 38
Winter Olympics *3:* 573
Wire *1:* 45; *2:* 382; *4:* 635; *6:* 1010, 1050, 1052
Wöhler, Friedrich *6:* 1120
Wolfram *6:* 1095
Wolframite *6:* 1095
Wollaston, William Hyde *5:* 845; *6:* 1032
Woodward, Robert Burns *5:* 817
Wool *6:* 1143
WordPerfect Corp. *2:* 295
WordStar *2:* 296
Workgroups *2:* 297
World Health Organization *6:* 1173
World War I *2:* 261
World War II *2:* 262; *5:* 925
World Watch Institute *1:* 10
World-Wide Standardized Seismograph Network (WWSSN) *5:* 918
WORM *2:* 305; *4:* 764
Wright, Almroth *5:* 825
Wright, James *5:* 927
Wright, Orville *1:* 30; *4:* 636; *6:* 1156
Wright, Wilbur *1:* 30; *4:* 636; *6:* 1156
Wrigley, William, Jr. *2:* 265
Wristwatch *2:* 275
Writing *1:* 43; *2:* 400

X

X chromosome *5:* 920
Xenon *3:* 539; *4:* 726; *5:* 883; *6:* **1159-1160**
Xerography *5:* 827
xocoatl *2:* 266
X-ray *2:* 282, 355, 385; *4:* 670, 722; *5:* 817, 876; *6:* 1007, 1108, 1111, 1112, 1118, **1160-1162,** 1163
X-ray astronomy *6:* **1162-1165**
X-ray crystallography *6:* 1007, 1162, 1166
X-ray fluoroscope *6:* 1162, 1166
X-ray machine *4:* 674; *6:* 1110, **1165-1167**
X-ray radiation *6:* 1108
X-ray spectroscopy *3:* 551
X-ray star *6:* 1162

Y

Yale, Linus, Jr. *4:* 659
Y chromosome *5:* 920
Yeager, Chuck *1:* 35
Year *2:* 271
Yeast *3:* 542; *6:* **1169-1171**
Yellow fever *2:* 336; *6:* **1171-1173**
Yellow journalism *2:* 284
Yew tree *2:* 264
Yogurt *6:* **1173-1174**
York *6:* 1081
Young's modulus *2:* 379
Young, Thomas *2:* 379
Yukawa, Hideki *6:* 1023

Z

Zamboni, Frank J. *3:* 572
Zeeman, Erik Christopher *6:* 1077
Zeppelin, Ferdinand von *1:* 29
Zero *5:* 883; *6:* **1175-1176**
Ziegler, Karl *5:* 843
Zinc *3:* 563, 565; *4:* 676, 758; *5:* 929; *6:* 1068, 1089, 1124
Zinjanthropus *3:* 559
Zipper *6:* 1126, **1176-1177**
Zoetrope *4:* 714
Zoopraxiscope *4:* 716
Zworykin, Vladimir *2:* 258; *6:* 1058
Zygote intrafallopian transfer *3:* 588

Boldfaced numbers indicate main entry pages; italicized numbers indicate volume number